Editors

Lecture Notes in Chemistry

Edited by G. Berthier M.J.S. Dewar H. Fischer
K. Fukui G.G. Hall H. Hartmann H.H. Jaffé J. Jortner
W. Kutzelnigg K. Ruedenberg J. Tomasi

40

A. Plonka

Time-Dependent Reactivity of Species in Condensed Media

Springer-Verlag
Berlin Heidelberg New York London Paris Tokyo

Author

A. Plonka
Technical University (Politechnika)
Institute for Applied Radiation Chemistry
93-590 Lódź, Wroblewskiego 15, Poland

ISBN-13: 978-3-540-16796-9 e-ISBN-13: 978-3-642-93326-4
DOI: 10.1007/978-3-642-93326-4

PREFACE

These notes on the use of one particular form of the time-dependent rate constant to describe the reaction patterns in condensed media have been put together primarily to encourage chemists to try and accept this new way of experimental data treatment. A number of applications is shown and interpretative aspects are discussed. Emphasized are the problems that need to be currently solved. Some of them are also of current interest in condensed phase physics from which the chemical kinetics benefits a great deal. It was inevitable that the choice of subject matter from both rapidly expanding fields and its form of presentation reflect to some extent the author's own interests and some important topics are treated briefly or even omitted. Fully recognizing this, I would like to acknowledge with gratitude the contributions to the subject of all my coworkers in the Laboratories of Lodz, Detroit Mi, Mülheim/Ruhr, and Houston Tx, and of those who helped me in preparing this text. Dr. Wlodzimierz Lefik and my son Wojciech recalculated most of the experimental results, Mrs. Aleksandra Karczewska redrew all the figures. Special thanks go to my wife Ewa for her invaluable assistance in all works and for the final form of the text.

AP

Lodz, February 1986

CONTENTS

1. INTRODUCTION

In the late seventies we have endeavoured to describe adequately (Plonka et al. 1979) the complex decay patterns of radiation produced hydrogen atoms in low-temperature aqueous glasses with the use of the time-dependent rate constant k(t) of the form

$$k(t) = Bt^{\alpha-1} , \qquad 0 < \alpha \leqslant 1 \tag{1.1}$$

where B and α are constants. This peculiar form (1.1) of the time-dependent rate constant was derived from the continuous time random walk, CTRW, model developed for the anomalous electron transit-time dispersion in amorphous solids (Scher and Montroll 1975) and used to describe the decay of excess electrons trapped in rigid glassy matrices (Hamill and Funabashi 1977).

The main feature of the decay patterns of trapped hydrogen atoms, adequately described with the use of the time-dependent rate constant (1.1), consists in the initial period of rapid decay followed by much slower decay. Similar behaviour is often observed in reactions of species trapped in condensed media, cf. the survey of experimental results in Chapters 2 - 5. Attempts to explain it in terms of initial nonrandom spatial distributions of the reacting species or initial distribution of the reacting species between shallow and deep traps were unsuccessful. For the extensively studied system of hydrogen atoms trapped in low-temperature aqueous glasses the first possibility was eliminated by relaxation measurements (Zimbrick and Kevan 1967; Ershov et al. 1968) showing that hydrogen atoms are distributed uniformly throughout the matrix. The second possibility was carefully analyzed and rejected because it was hard to accept the dramatic changes of the initial distribution of hydrogen atoms between deep and shallow traps with minor changes in experimental conditions (Sprague and Schulte-Frohlinde 1973). It was also shown (ibid.) that the experimental results are hard to explain in terms of the classical theory of diffusion controlled reaction.

Time dependence of the form (1.1) is quite different from the variation with time of the rate constant for bimolecular reaction given by the classical diffusion theory (Smoluchowski 1917; Waite 1960)

$$k(t) = k_1\left[1 + (\pi\tau)^{-1/2}\right] \tag{1.2}$$

with $k_1 = 4\pi r_o D$ and $\tau = Dt/r_o^2$, where D denotes the sum of diffusion coefficients of the reacting particles and r_o is the capture radius. The difference is that for $t \rightarrow \infty$ k(t) according to eqn. (1.2) attains the constant value of $4\pi r_o D$ while k(t) according to eqn. (1.1) decreases monotonically to zero with the rate controlled by the numerical value of α. The classical time-independent rate constant corresponds to $\alpha = 1$.

Successful use of the rate constant (1.1) to describe the complex decay patterns of trapped hydrogen atoms in low-temperature glasses posed two questions. The first one was the physical interpretation of the constants α and B in eqn. (1.1), the second one was concerned with the reaction mechanism. As there was no H/D kinetic isotope effect in the decay of trapped hydrogen atoms in the temperature region around 77 K in which most of the experiments were performed (Kroh and Plonka 1975; Bøckmann and Moan 1976), the essence of the CTRW model for electron transit-time dispersion, the hopping mechanism was eliminated. It was suggested (Kroh and Plonka 1975) that the trapped species become mobile and/or reactive due to trap destruction caused by matrix relaxation. Attention was paid (Plonka et al. 1982) to the formal identity of the relaxation function (Kohlrausch R 1854; Kohlrausch F 1863)

$$\Phi(t) = \exp\left[-(t/\tau_0)^{\alpha}\right] \qquad (1.3)$$

with the first-order reaction kinetics with a time-dependent rate constant of the form (1.1), i.e.

$$c/c_0 = \exp(-Bt^{\alpha}/\alpha) \; . \qquad (1.4)$$

Kohlrausch relaxation function (1.3) used over the years to describe the nonexponential decays in a variety of relaxation phenomena, can be regarded (cf. Majumdar 1971) as the result of superposition of many simple exponential decays. Thus it was natural to attempt to analyze eqn. (1.4) in the same way, i.e. in terms of a continuous distribution of exponential processes. Formally this is accomplished by introducing the probability density $f(\tau)$

$$\exp\left[-(t/\tau_0)^{\alpha}\right] = \int_0^{\infty} f(\tau)\exp(-t/\tau)\,d\tau \qquad (1.5)$$

where $\tau_0 = (\alpha/B)^{1/\alpha}$ is the effective lifetime. Then, by definition, $\exp\left[-(t/\tau_0)^{\alpha}\right]$ is the Laplace transform of the lifetime density $f(\tau)$ so that the latter is the inverse Laplace transform of $\exp\left[-(t/\tau_0)^{\alpha}\right]$. For large τ, which are of interest in kinetics, the inverse Laplace transform can be evaluated by saddle point calculation. The result is (Majumdar 1971)

$$f(\tau) \sim \frac{(\alpha\tau)^{(2-\alpha)/2(1-\alpha)}}{\tau^2\left[2\pi\alpha(1-\alpha)\right]^{1/2}\tau_0^{\alpha/2(1-\alpha)}} \cdot \exp\left[-(1-\alpha)(\alpha\tau/\tau_0)^{\alpha/(1-\alpha)}\right] \qquad (1.6)$$

There are, cf. Appendix B, analytical solutions for $\alpha = 1/3$ and $1/2$, and series expansions to evaluate $f(\tau)$ with accuracy adequate to experimental results. Here, cf. Fig. 1.1, some distributions are depicted to illustrate their dependence on the numerical value of α . Thus the

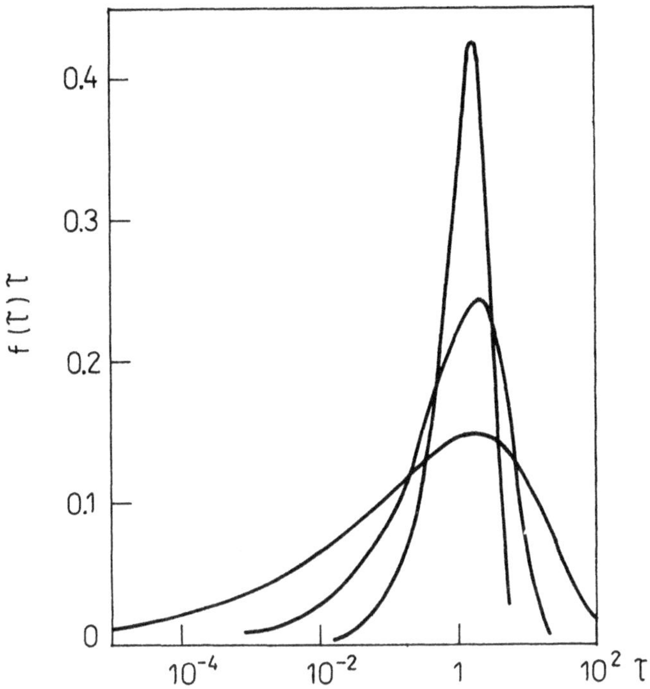

Fig. 1.1. Reactivity distributions from inverse Laplace transform of $\exp[-(t/\tau_o)^\alpha]$ for α = 1/3, 1/2, and 2/3 (from bottom to top)

numerical value of α determines the width of the density distribution while B or τ_o determines the position of the distribution on the time scale. For α = 1 we have delta distribution and in this case B is equal to the classical time-independent rate constant.

Making use of Arrhenius equation

$$\tau = s\exp(\Delta E/RT) \tag{1.7}$$

it is straightforward to pass from reactivity distribution to the distribution of activation energies

$$g(E) = f(\tau)(d\tau/dE) = f(\tau)\tau/RT\ . \tag{1.8}$$

It is hard to derive the distribution of activation energy from the second-order kinetic equation with time-dependent rate constant (1.1) which reads, cf. Appendix A,

$$c^{-1} - c_o^{-1} = Bt^\alpha/\alpha \tag{1.9}$$

for an equal concentration case, and

$$(c_A/c_B)/(c_{A,o}/c_{B,o}) = \exp\left[-(c_A - c_B)Bt^{\alpha}/\alpha\right] \quad (1.10)$$

for a general case. It will be shown, however, in Chapter 6 that the barrier dispersion model (Bendler and Shlesinger 1985) fully supports the conclusion which follows from the above analysis of first-order processes:

> kinetic equations with time-dependent rate constant reflect the distribution of activation energies for reactions of trapped species in condensed media.

With that in mind we will be less surprised to witness the widespread application of the time-dependent rate constant (1.1) in the last few years to describe adequately the reactivities of substrates as diverse as electrons and macroradicals or in media so different as glasses and micelles.

If we take the distribution of activation energy as the only necessary condition for application of the time-dependent rate constant (1.1), it will be not only easy to accept the universality of the presented kinetic treatment but also to admit that these is no unique mechanism behind this type of kinetics. Some of them, however, will be seen to operate quite often.

It is a happy situation that the development of the presented kinetics is paralleled by the widespread use of Kohlrausch relaxation function (1.3) in diverse areas of condensed phases physics, as the physical origin of the nonexponential relaxation is discussed from several different points of view including non-Markovian stochastic processes and basic quantum mechanisms (Rajagopal et al. 1983). The Kohlrausch relaxation function (1.3) is now recognized as the characteristic function of the stable or Lévy distributions and some suggestions are made about the physical models that might lead to it (Montroll and Bendler 1984; Bendler 1984). All this gives us a deeper insight into the molecular origins of reactivity distributions in condensed media, cf. Chapter 6.

REFERENCES

Bendler JT (1984) Lévy (stable) probability densities and mechanical relaxation in solid polymers. J Statist Phys 34: 625-637

Bendler JT, Shlesinger MF (1985) Derivation of the Kohlrausch-Williams/Watts decay law from activation energy dispersion. Macromolecules 18: 591-921

Bøckmann B, Moan J (1976) On the decay of H and D atoms in aqueous sulphuric acid glases. Isotope effects. Chem Phys Lett 42: 575-577

Ershov BG, Chernova GP, Grinberg OYa, Lebedev YaS (1968) Spatial distribution of radicals in γ-irradiated frozen aqueous solutions (Russ). Izv Akad Nauk SSSR, Ser Khim 2439-2443

Hamill WH, Funabashi K (1977) Kinetics of electron trapping reactions in amorphous solids; a non-Gaussian diffusion model. Phys Rev B 16: 5523-5527

Kohlrausch F (1863) Ueber die elastische Nachwirkung bei der Torsion. Pogg Ann 119: 337-368

Kohlrausch R (1854) Theorie des elektrischen Rückstandes in der Leidener Flasche. Pogg Ann 91: 179-214

Kroh J, Plonka A (1975) Trapped hydrogen atom decay in γ-irradiated sulfuric acid glasses at 63 - 90 K. J Phys Chem 79: 2600-2602

Majumdar CK (1971) Stress relaxation function of glass. Solid State Comm 9: 1087-1090

Montroll EW, Bendler JT, (1984) On Lévy (or stable) distributions and the Williams-Watts model of dielectric relaxation. J Statist Phys 34: 129-162

Plonka A, Bogus W, Kroh J (1982) Effect of irradiation temperature on the decay rate of trapped hydrogen atoms in sulfuric acid glasses. Radiat Phys Chem 19: 389-392

Plonka A, Kroh J, Lefik W, Bogus W (1979) Kinetics of trap-limited hydrogen atom decay in γ-irradiated sulfuric acid glasses. J Phys Chem 83: 1807-1810

Rajagopal K, Ngai KL, Rendell RW, Teitler S (1983) Nonexponential decay in relaxation phenomena. J Statist Phys 30: 285-292

Scher H, Montroll EW (1975) Anomalous transit-time dispersion in amorphous solids. Phys Rev B 12: 2455-2477

Smoluchowski M (1917) Mathematical theory of the kinetics of the coagulation of colloidal solutions. Z physik Chem 92: 129-168

Sprague ED, Schulte-Frohlinde D (1973) Electron spin resonance investigations of the disappearance of trapped hydrogen atoms in γ-irradiated sulfuric acid glasses. J Phys Chem 77: 1222-1225

Waite TR (1960) Bimolecular reaction rates in solids and liquids. J Chem Phys 32: 21-23

Zimbrick J, Kevan L (1967) Paramagnetic relaxation of trapped hydrogen atoms in irradiated frozen aqueous solutions. J Chem Phys 47: 5000-5008

2. REACTION KINETICS OF SPECIES TRAPPED IN GLASSY MATRICES

Even the most reactive intermediates can be trapped following irradiation of proper matrices at sufficiently low temperatures. Glassy matrices convenient for optical studies of trapped intermediates were always of the first choice and they are the first to be discussed.

EXCESS ELECTRONS

Reactivity of electrons in disordered media is one of the fundamental problems in the chemistry and physics of condensed phase. It is particularly relevant to a variety of chemical areas including charge conduction in dielectrics, electron transfer in condensed phases, and of course radiation chemistry.

Post-Irradiation Decay

Excess electrons trapped in rigid glassy matrices do react slowly with electron-accepting impurities. No simple conventional rate law is obeyed: successive half-lives increase greatly and the rates are often independent of temperature over a wide range. These peculiar modes of electron decay have been described quite successfully by the tunnelling mechanism which invokes direct electron-transfer processes between acceptors and initial solvent traps (Kroh and Stradowski 1975).

In 1977, when the tunnelling models seemed to be generally accepted, Hamill and Funabashi reported that a certain class of decay patterns for trapped electrons in aqueous glasses can be described by the pseudo first-order kinetic equation

$$-dn(t)/dt = k(t)\, n(t)\, c_s(t) \qquad (2.1)$$

with the time-dependent rate constant of the form

$$k(t) = Dt^{\alpha-1}, \qquad 0 < \alpha \leq 1 \qquad (2.2)$$

i.e. by

$$n(t)/n(0) = \exp(-c_s B t^{\alpha}/\alpha) \qquad (2.3)$$

where the concentration of acceptor, c_s, far in excess of that of the electrons, was assumed to be constant.

Furthermore, the particular time-dependent form of the rate constant, eqn. (2.2), was shown (Hamill and Funabashi 1977) to be consistent with the long-tail jump time distribution function for a random walk in disordered materials, cf. Chapter 6. To illustrate the applicability of eqn. (2.3) Hamill and Funabashi have used the experimental data of Miller

(1975). Equating the ratio n(t)/n(0) in eqn. (2.3) to the ratio of normalized optical absorbances they replotted Miller data for electron decay in the range 10^{-6} - 10^{3} s at 77 K for 6 M aqueous NaOH containing acceptors like BrO_3^-, Fig. 2.1, Co(ethylenediamine)$_3^{3+}$, Fig. 2.2, $Co(CN)_6^{3-}$ and Np^{7+}, Fig. 2.3, in the coordinate system ln ln(A^o/A) vs. ln t in which eqn. (2.3) yields a straight line. The experimental data are seen to fit fairly well eqn. (2.3) with (B/α), taken as the measure of relative efficiency of acceptors, equal to 0.1, 0.5, 0.4 and 1 for BrO_3^-, Co(ethylenediamine)$_3^{3+}$, $Co(CN)_6^{3-}$ and Np^{7+}, resp. These small changes of the efficiency factor, within one order of magnitude, are to be compared with those required in the curve-fitting process for tunnelling. In terms of the tunnelling model BrO_3^- was considered to be abnormal and in an efficiency factor of 10^{-11} had to assigned, Co(ethylenediamine)$_3^{3+}$ and $Co(CN)_6^{3-}$ were normal with an efficiency factor of unity, but for Np^{7+} the efficiency factor was 10^4 (Miller 1975).

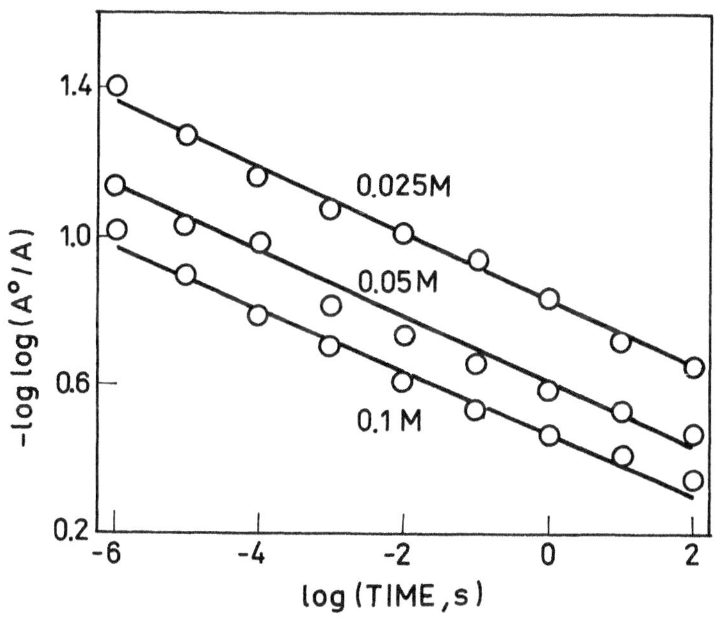

Fig. 2.1. Decay at 77 K of the normalized optical absorbance at 550 nm of trapped electrons in aqueous 6 M NaOH containing 0.025 - 0.1 M of $NaBrO_3$. All slopes are nearly equal, α = 0.087

Because of the above mentioned consistency of the particular time-dependent form of the rate constant, eqn. (2.2), with the long-tail jump time distribution function for a random walk in disordered materials, further theoretical treatments of electron decay reactions evolved around two distinct, limiting assumptions. The first was that trapped electrons react directly with scavengers, the second that electrons hop from one trap to another until encountering the scavenger with which

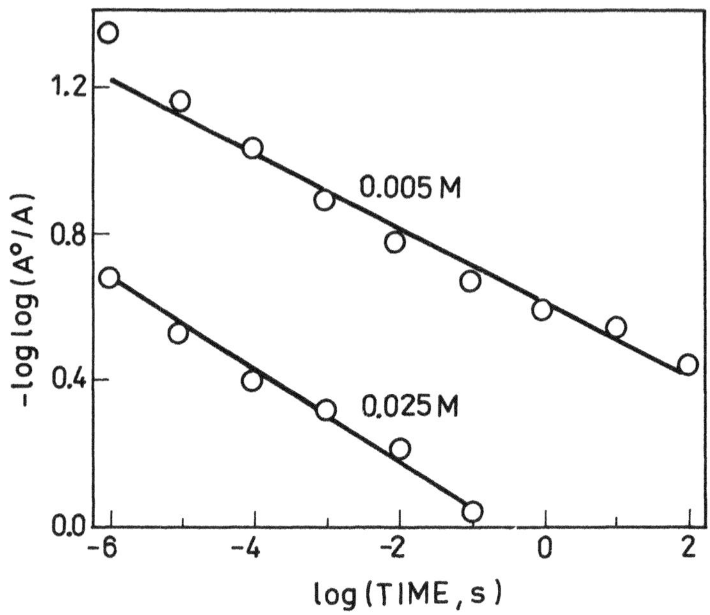

Fig. 2.2. Decay at 77 K of the normalized optical absorbance at 550 nm of trapped electrons in aqueous 6 M NaOH containing Co(ethylenediamine)$_3^{3+}$ α = 0.100 (0.0025 M) and α = 0.080 (0.005 M)

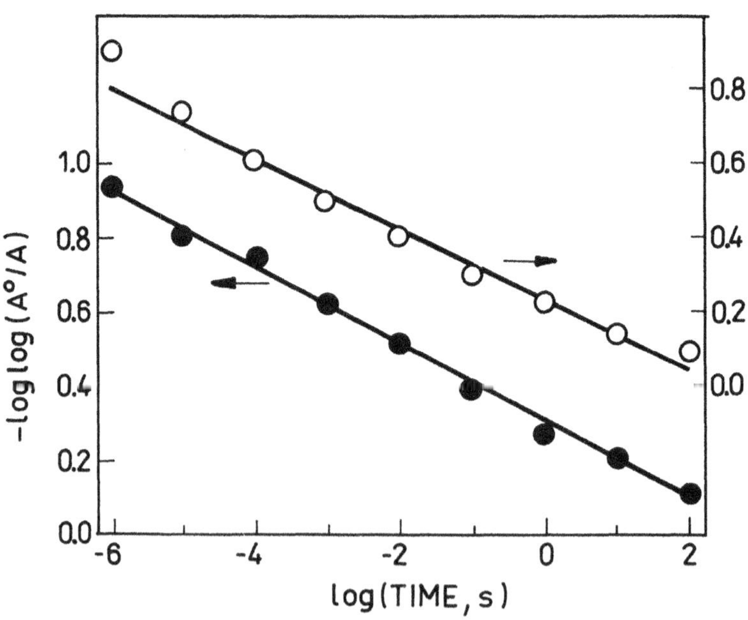

Fig. 2.3. Decay at 77 K of the normalized optical absorbance at 550 nm of trapped electrons in aqueous 6 M NaOH containing 0.025 M of $Co(CN)_6^{3-}$ (○), α = 0.096, and $7.7 \cdot 10^{-3}$ M of Np^{7+} (●), α = 0.103

they will react. Much effort has been devoted to the controversial issue of finding out the correct model. It had to be agreed that both the direct tunnelling model and the trap-to-trap hopping can fit the experimental data equally well (van Leeuwen et al. 1980). There is a fundamental reason for this. The time-dependent rate constant

$$k(t) = -\left[dn(t)/dt\right]/n(t) \qquad (2.4)$$

calculated from n(t) given by the tunnelling models is adequately approximated by eqn. (2.2) (Lefik et al. 1980), as illustrated in Fig. 2.4 for the model of Bartczak et al. (1977).

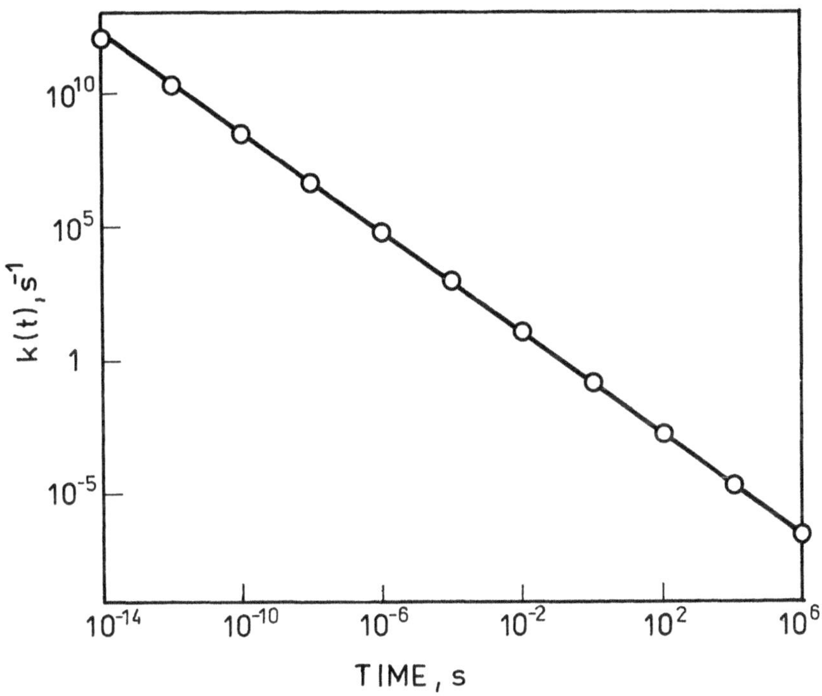

Fig. 2.4. Time dependence of rate constant calculated according to eqn. (2.4) from n(t) given by the tunnelling model of Bartczak et al. (1977) for reactions of electrons with NO_3^- at concn. $5\cdot10^{-2}$ M

It is seen in Fig. 2.4 that for time range $10^{-12} - 10^{6}$ s the time-dependent rate constant is quite well approximated by eqn. (2.2), viz.

$$k(t) = 0.174t^{(0.065-1)} \qquad (2.5)$$

The numerical value of $\alpha = 0.065$ is close enough to those observed experimentally, cf. Figs. 2.1, 2.2, 2.3.

In the tunnelling model of Bartczak et al. (1977) due to the distribution of trapped electron energies and the distribution of electron-acceptor distances there is a distribution of electron reaction rates with the density f(k) depicted in Fig. 2.5 in the coordinate system f(k)k vs. k. It is compared with that obtained from inverse Laplace transform of the right-hand side of eqn. (2.3) which for this special case of small values of α, cf. Appendix B, is given by

$$f(k)k = (B/k^{\alpha}) \exp(-B/k^{\alpha}) \qquad (2.6)$$

There are differences for very fast processes which cease rapidly with time, and for times longer than 10^{-8} s the normalized distributions

$$f(k,t)k = f(k)k \exp(-kt) \qquad (2.7)$$

are indistinguishable.

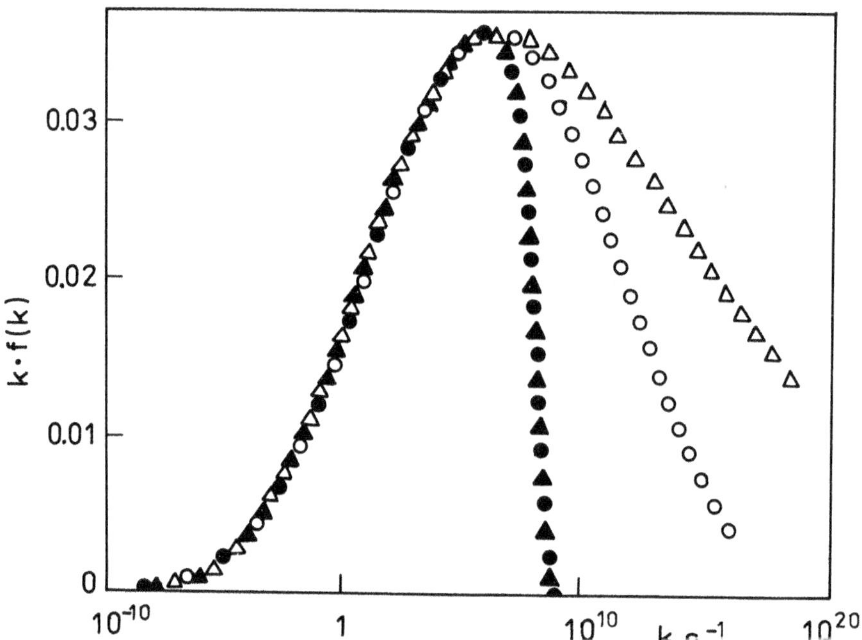

Fig. 2.5. Comparison of the distributions of reactivities: initial distributions from the tunnelling model (○) and from eqn. (2.6) (△); normalized distribution after 10^{-8} s from tunnelling model (●) and from eqn. (2.6) (▲)

The presented example, cf. also Dainton et al. (1975), explains not only why both the direct tunnelling model and trap-to-trap tunnelling model were found to fit the experimental data equally well, but also makes useless the efforts to solve the controversial issue of finding out the correct model by analysis of kinetic data. In the author's opinion there are a number of mechanisms leading to eqn. (2.2) and there is hope that estimation of reactivity distribution from the inverse Laplace transforms will always match those predicted by the correct model. This seems to be the main reason of the universality of the discussed kinetics.

Spectral Relaxation

According to Funabashi and Hamill (1979) the time dependence of optical absorption spectra for localized electrons in glasses involves two related problems. There are often at least two spectral regions, visible and infrared, for the optical response of the localized electrons. One has to answer the question, does the spectral difference represent two distinct localized states or reflect two distinguishable stages of a continuous dielectric relaxation process around the localized charge. The answer is not of primary importance for the kinetics of spectral relaxation but it is worth to note that Buxton et al. (1976) demonstrated for some aqueous glasses that the infrared and visible electrons belong to chemically distinguishable states. There is difference in their reactivities with electron scavengers, the fully solvated electrons being more reactive. Accepting the distinguishable states, the second problem is the transport mechanism. Contrary to the decay mechanism which was commonly accepted among radiation chemists that the electron tunnels between the initial trap site and acceptor, Funabashi and Hamill (1979) preferred an alternative decay mode based on a CTRW model, cf. above. Assuming that for the conversion of infrared, or trapped electrons in their nomenclature, into visible or solvated electrons

$$e_t^- \longrightarrow e_s^- \tag{2.8}$$

the deep trap simulate the role of the scavenger the spectral relaxation kinetics should follow eqn. (2.3). It was shown to be the case in a wide range of systems. Here are some examples.

Funabashi and Hamill (1979) reexamined the data of Miller et al. (1976) for several undoped alcohols at 77 K in the range 10^{-7} to 10^{2} s after pulsed irradiation. At the longest wavelength, 950 nm, the decay does not follow eqn. (2.3), probably more than a single species of e_t^- is involved. The absorbance growth at 550 nm for ethanol was examined because only one spectral component was expected. The results are depicted in Fig. 2.6.

Fig. 2.6 includes the data of Klassen et al. (1975) on spectral shifts in ethanol at 76 K. Their results for decay of e_t^- at 1400 nm were not well described by eqn. (2.3) when A^o was taken as the measured value from their Fig. 3. For the yield at 200 ns after the onset of a 140 ns pulse, $G\epsilon = 3.14 \cdot 10^4$, where G is the number of electrons per 100 eV and ϵ is the extinction coefficient in $M^{-1}cm^{-1}$. Some improvement was found for $G\epsilon = 4.2 \cdot 10^4$ at time zero, cf. the interpolated data included in Fig. 2.6. The slope, $\alpha = 0.20$, agrees well with that for the results of Miller et al. (1976), $\alpha = 0.18$.

Ogasawara and Kevan (1978) followed electron solvation kinetics in ethanol containing 0.1 M NaOH from 83 to 123 K by ruby laser photoionization of $5 \cdot 10^{-4}$ M β-naphtolate. The results for growth of solvated electron

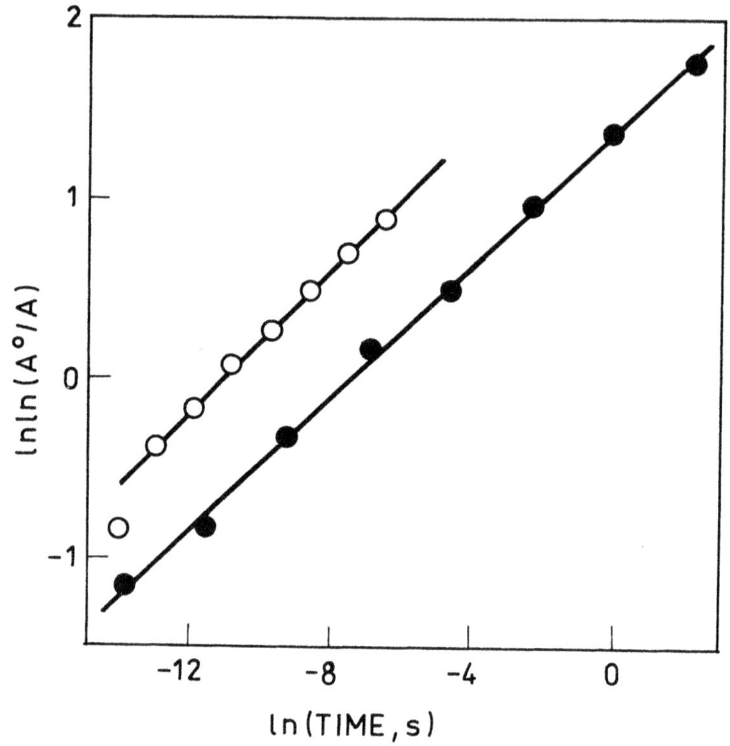

Fig. 2.6. The growth of optical absorbance of e_s^- in ethanol at 550 nm (●) from Miller et al. (1976) and the decay of optical absorbance of e_t^- in ethanol at 1400 nm (○) from Klassen et al. (1975)

absorption at 580 nm following 20 ns ruby laser pulses were analyzed by Funabashi and Hamill (1979), cf. Fig. 2.7. For A_∞ = 0.196, at 108.0, 98.0, 92.8, 97.7, and 83.1 K (from top to bottom) the slopoes are 0.134, 0.104, 0.099, 0.081, and 0.062. Although there are no data for 77 K one may expect α to be very much less than the values from Fig. 2.6. The difference was attributed to the presence of 0.1 M NaOH and an increased dispersion of trap depths.

As the last example the experimental data of Baxendale and Sharpe (1976) on electron transfer in 1-propanol at 108 K are presented in Fig. 2.8. For the growth in the visible region the measured values of the optical absorbance of e_s^- for the final state have been converted to the equivalent values of the absorbance of e_t^- for the initial state, cf. Funabashi and Hamill (1978).

Thus there is no doubt that both spectral relaxation and electron scavenging, discussed in the previous section, in a wide range of systems follow eqn. (2.3). However, it does not mean that the transport mechanism based on CTRW model is proved. In spectral relaxation of trapped electrons

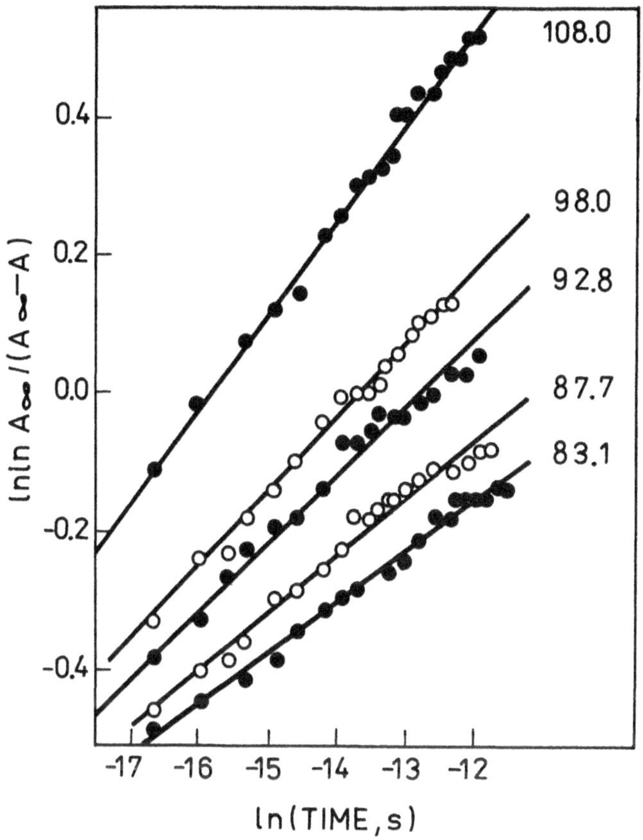

Fig.2.7. The growth of optical absorbance of e_s^- at 550 nm in ethanol containing 0.1 M NaOH following 20 ns ruby laser photoionization of $5 \cdot 10^{-4}$ M β-naphtolate at 83 - 123 K

the alternative that the spectral difference reflect two distinguishable stages of a continuous dielectric relaxation process around the localized charge cannot be exluded. Especially as there is shown the example that transient dielectric constant follows eqn. (2.3) (Hamill 1981), cf. Fig. 2.9.

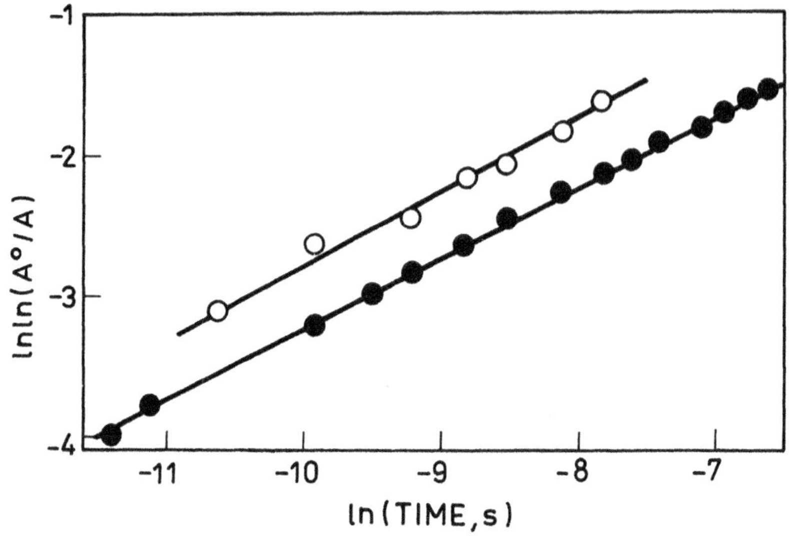

Fig. 2.8. Electron transfer in 1-propanol at 108 K following pulse irradiation from shallow traps to deep traps based on absorbances at 1300 nm (●) and 550 nm (○)

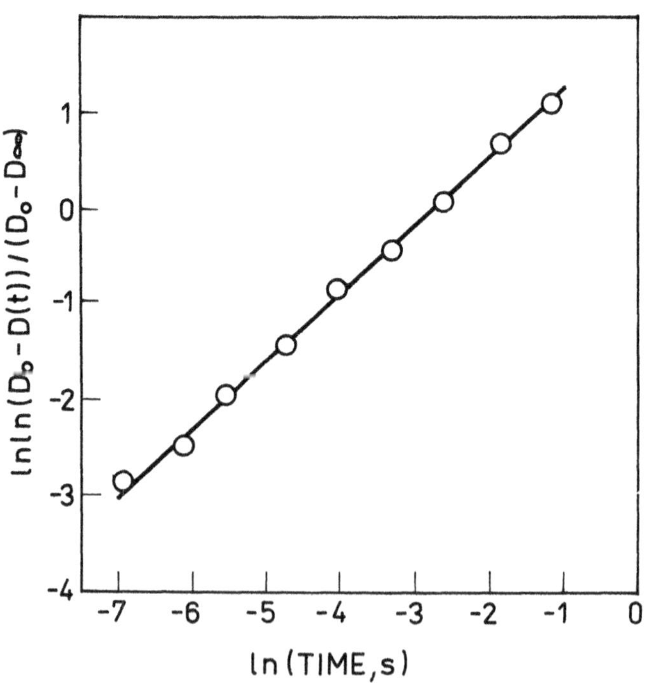

Fig.2.9. Transient dielectric constant D(t) in glycerol at 202 K

Photostimulated Decay

Photostimulated decay of trapped electrons in alkaline ice can be included into the above discussed class of decay patterns (Plonka et al. 1979). Photobleaching of trapped electrons, e_t^-, in alkaline ice consists of promotion of electrons to a mobile state, e_m^-, from which they can be scavenged or redistributed in new trapping sites, cf. (Rice and Kevan 1977). Thus

$$e_t^- \xrightarrow{h\nu} e_m^- + T \; , \tag{2.9}$$

$$e_m^- + S \longrightarrow P \; , \tag{2.10}$$

$$e_m^- + T \longrightarrow e_t^- \; , \tag{2.11}$$

where T denotes the trapping site, S and P are the scavenger and product of scavenging, respectively.

In the above pattern the rate of trapped electron decay is equal to the rate of product formation in reaction (2.10)

$$-d\left[e_t^-\right]/dt = d\left[P\right]/dt \; , \tag{2.12}$$

which we assume to be given by

$$d\left[P\right]/dt = k(t)\left[e_{m,s}^-\right]\left[S\right] \; , \tag{2.13}$$

where k(t) denotes the time-dependent scavenging rate constant of the form given by eqn. (2.2), and the subscript s was added to mark the fraction of mobile electrons which are reactive towards the scavenger.

Because of very small quantum yields of photobleaching (Rice and Kevan 1977), there is always even in alkaline ice without deliberately added scavenger

$$\left[e_{m,s}^-\right] \ll \left[S\right] \; , \tag{2.14}$$

and pseudo first-order reaction kinetics with time-dependent rate constants should be observed for product formation in reaction (2.10). It is, however, extremely difficult to follow by the ESR method the accumulation of products in reaction (2.10). Instead it is relatively easy to follow the decay of trapped electrons, which because of eqn. (2.12), should also decay according to the first-order kinetics formally given by

$$-d\left[e_t^-\right]/dt = k_{eff}(t)\left[e_t^-\right] \; , \tag{2.15}$$

where $k_{eff}(t)$ denotes the time-dependent effective rate constant depending, cf. eqn. (2.13) and reaction (2.9), evidently on intensity of light and the concentration of the scavenger added.

Integrating eqn. (2.15) with $k_{eff}(t)$ in the form given by eqn. (2.2) one obtains

$$\ln([e_t^-]_o/[e_t^-]) = (B'/\alpha)t^{\alpha} . \qquad (2.16)$$

An example of the obtained data for trapped electron photobleaching in 8 M NaOH glass without and with added scavengers is presented in Fig. 2.10. The slopes of the lines, α values equal to 0.62, 0.65, and 0.61 for 8 M NaOH glass without and with 10^{-3} and 10^{-2} M of $K_3Fe(CN)_6$ added, coincide within the estimated standard deviation of α values equal to about 5 per cent. Numerical values of B' are equal to 0.023, 0.027, and 0.063, resp. According to eqns. (2.6) and (2.16) some additivity is expected and it really holds. Increase of B' values in the presence of 10^{-3} and 10^{-2} M of scavenger is equal to 0.004 and 0.040, resp.

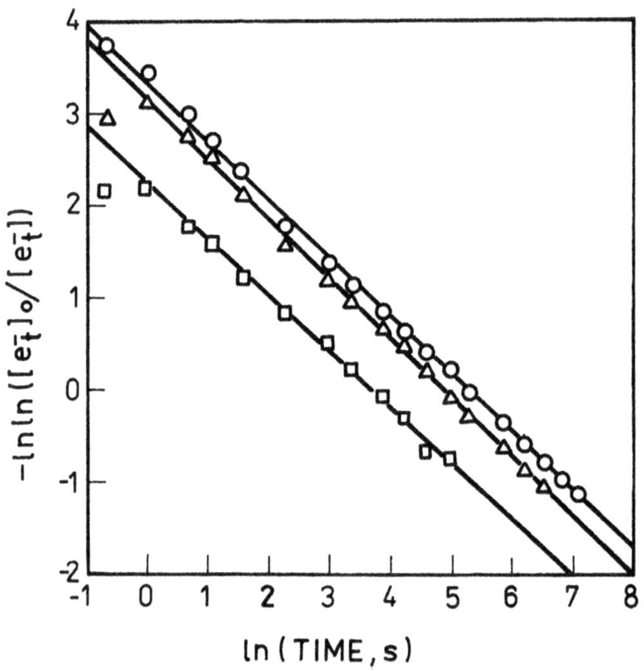

Fig. 2.10. Decay of electrons trapped in 8 M NaOH glass without (○) and with 10^{-3} (△) and 10^{-2} M (□) of $K_3Fe(CN)_6$ added upon photobleaching with wide band visible light. The estimated radiation yield of trapped electrons before photobleaching were equal to 1.9, 1.4, and 1.2 per 100 eV, respectively

Concluding that the photostimulated decay of trapped electrons in alkaline ice belongs to the class of decay patterns described by the time-dependent scavenging rate constant of the form given by expression (2.2), it is worth mentioning one point. The numerical values of α for the photostimulated electron decay in alkaline ice, 0.61 - 0.65, are much greater than those, equal to about 0.1, estimated by Hamill and Funabashi (1977) from the results reported by Miller (1975) for spontaneous electron decay in the time range 10^{-6} - 10^{3} s. This indicates that photoassisted electron transport is less dispersive than the spontaneous transport.

Photostimulated Conversion into Trapped Hydrogen Atoms

The photostimulated conversion of radiation produced trapped electrons into trapped hydrogen atoms was observed in neutral (Konigshang et al. 1977; Plonka and Kowalski 1977) and alkaline glasses (Konigshang et al. 1977). As an example, the neutral LiCl glass will be discussed (Plonka et al. 1981).

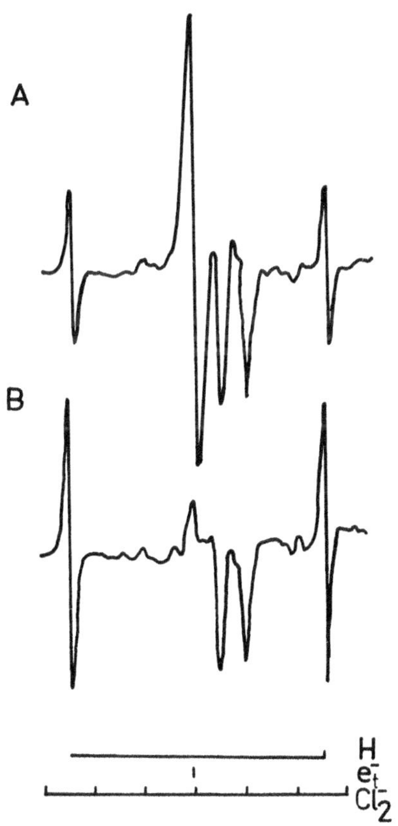

Fig. 2.11. X-band (0.3 mW) ESR spectra of 15 M LiCl glass γ-irradiated at 77 K before (A) and after (B) photobleaching with wide band visible light at 77 K

The ESR spectrum recorded at 77 K, at relatively low microwave power, of 15 M LiCl glass γ-irradiated at 77 K in the dark is presented in Fig. 2.11 A. It consists, cf. (Plonka and Kowalski 1977) of the doublet split about 505 G corresponding to the trapped hydrogen atoms H_t, the singlet at $g \sim 2.002$ with width of about 15 G corresponding to the trapped electrons, e_t^-, and poorly resolved at low microwave power seven line spectrum at $g = 2.0038$ with hyperfine splitting of about 100 G corresponding to the anion-radicals, Cl_2^-. Photobleaching at 77 K of 15 M LiCl glass γ-irradiated at 77 K results, cf. Fig.2.11 B, in the decay of trapped electrons with simultaneous decrease of Cl_2^- content and increase of H_t content. After complete removal of trapped electrons there is no further change in the ESR spectrum upon prolonged photobleaching with the wide band visible light. This indicates that part of the photostimulated electrons do react with the anion-radicals Cl_2^- and part do convert into trapped hydrogen atoms.

The photostimulated decay of trapped electrons in lithium chloride glasses, like in alkaline glasses, can be adequately described by pseudo first-order kinetics with the time-dependent scavenging rate constant of the form (2.2) depending on the light intensity and the concentration of the scavenger added.

Assuming that photomobilized electrons react in part with the anion-radicals

$$e_m^- + Cl_2^- \longrightarrow 2\ Cl^- \qquad (2.17)$$

and in part do convert into trapped hydrogen atoms in reaction

$$e_m^- + H_2O \longrightarrow H + OH^- \qquad (2.18)$$

one would expect the linear decrease of anion-radical Cl_2^- content and linear increase of trapped hydrogen content with the amount of photobleached electron. This assumption turns out to be true for anion-radicals Cl_2^- and false for H_t, cf. Fig. 2.12.

About 65 per cent of the photomobilized electrons do react with anion-radicals Cl_2^- according to reaction (2.17). To keep valid reaction (2.18) one would be forced to accept that in the presented example the effectiveness of electron conversion into trapped hydrogen atoms increases with the time of photobleaching from about 2 per cent at the beginning to about 13 per cent at the end.

Not seeing any reason to accept this let us recall the picture emerging from time-resolved photobleaching of trapped electrons in LiCl glasses (Walker and May 1974). These glasses seem to be bleached across the entire band, with rather low efficiency, to give a transient state which absorbs significantly in the near IR. This state, suggested to be an excited state associated in some way with halide ions, decays to reform the ground state e_t^- with a rate constant of about $5 \cdot 10^6\ s^{-1}$ at 77 K. Because of the low quantum efficiency of permanent bleaching the trapped electrons are to be transferred many times to this state to become mobile and/or reactive, i.e. to leave the trap. Each time the trapped

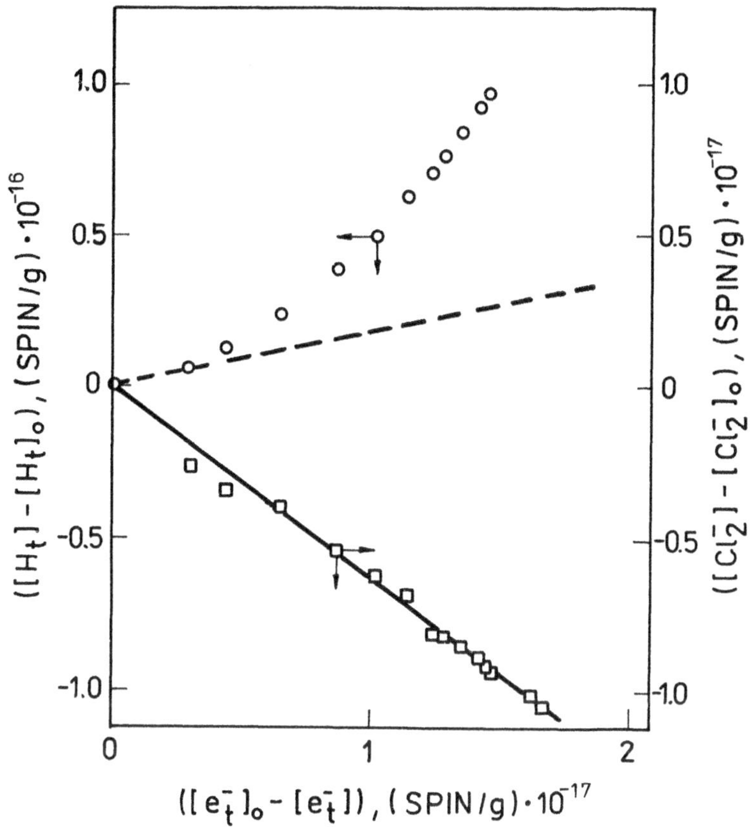

Fig. 2.12. Decrease of anion-radicals Cl_2^- content (□) and increase of trapped hydrogen atom content (○) vs. the amount of photobleached electrons in 15 M LiCl glass

electrons are excited, the water molecules forming the trap are forced to change their arrangement. Some water molecule configurations, unstable from the point of view of the electron trap, could favour the reaction of electron with water molecules. The number of such events should be strictly proportional to the number of returns from the excited state which in turn, under constant intensity of bleaching light, is proportional to the concentration of trapped electrons.

Basing on these arguments the rate of trapped hydrogen content increase should be equal to

$$d[H_t]/dt = \varkappa[e_t^-] , \qquad (2.19)$$

where $\varkappa$ denotes the constant depending like k_{eff} in eqn. (2.15), on the light intensity and the concentration of the scavenger added.

The concentration of trapped electrons decreases with the time of photobleaching according to eqn. (2.16). Introducing eqn. (2.16) into the rate equation (2.19) one gets

$$d[H_t]/dt = \varkappa [e_t^-]_o \exp(-Bt^{\alpha}/\alpha) \;, \tag{2.20}$$

and further

$$[H_t] - [H_t]_o = \varkappa [e_t^-]_o \int_0^t \exp(-Bt^{\alpha}/\alpha)dt \;, \tag{2.21}$$

where the right-hand side integral must be calculated numerically. In doing so, one obtains the perfect fit of the experimental data to relation (2.21), cf. Fig. 2.13.

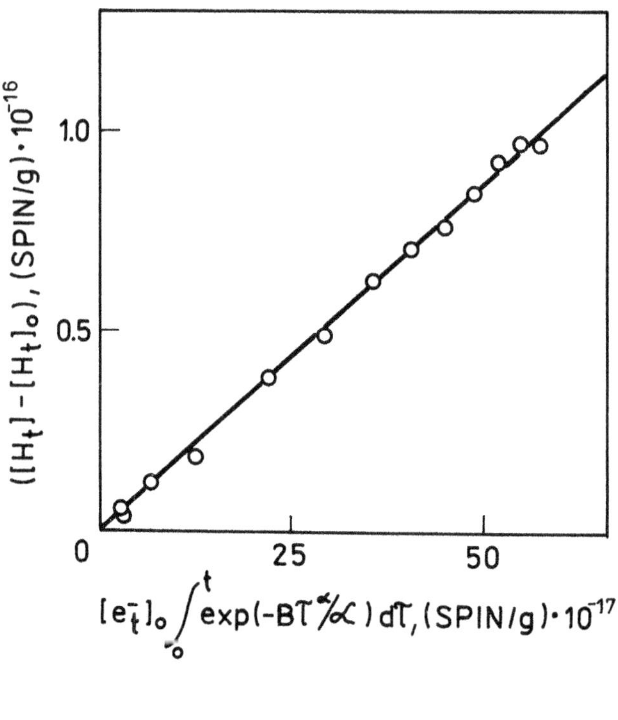

Fig. 2.13. Experimental verification of relation (2.21)

In repeated experiments the ratio of hydrogen atom content increase after complete photobleaching to the initial trapped electron concentration, $([H_t]_\infty - [H_t]_o)/[e_t^-]_o$, was constant and equal to about 6 - 7 per cent. It did not change for 7.5, 9.0 and 12 M LiCl glasses in which the initial concentration of trapped electrons decreases markedly with solute concentration (Gillis et al. 1978). To reduce this ratio to about 4 per cent one needs to add 0.5 - 1.0 M of isopropanol which is known to act as a very efficient scavenger of hydrogen atoms in glasses.

The addition of 10^{-3} M of $K_3Fe(CN)_6$, which acts as an electron scavenger, stops the increase of trapped hydrogen atom content during photobleaching. All this fully supports the presented kinetic picture of trapped electron conversion into trapped hydrogen atoms upon photobleaching.

Radioluminescence Kinetics

In a wide range of systems the kinetic behaviour of trapped species, mainly electrons, is followed by radioluminescence measurements which were shown (Plonka et al. 1984) to follow the kinetic pattern deduced from relation (2.2).

There are reasons to believe that the isothermal luminescence observed at the irradiation temperature and the first peak of the thermoluminescence glow curve are of the same origin (Hagekyriakou and Fleming 1982). In the case of γ-irradiated frozen solutions of LiCl discussed below both are most probably due to so-called infrared electrons (Klassen et al. 1980), i.e. electrons trapped in a very shallow cationic traps (Bartczak et al. 1981). Hence the intensity of light I emitted, at a constant temperature T, at time t after the irradiation for time τ, is proportional to the rate of escape of trapped electrons

$$I(\tau,t) = -\varkappa \frac{dn(\tau,t)}{dt} \qquad (2.22)$$

where $\varkappa$ denotes the proportionality factor. The rate of trapped electron escape is equal to

$$-\frac{dn(\tau,t)}{dt} = k(T,t)\ n(\tau,t) \qquad (2.23)$$

where k(T,t) denotes the time-dependent rate constant. Introducing the form (2.2) into eqn. (2.23) and integrating one obtains the relation

$$n(\tau,t) = n(\tau,0)\exp(-Bt^{\alpha}/\alpha)\ , \qquad (2.24)$$

which was shown, cf. eqn. (2.3) above, to describe adequately the decay of trapped electrons in frozen aqueous solutions both in the absence of light and during photobleaching.

Now, making use of eqns. (2.2), (2.22) - (2.24) one obtains

$$I(\tau,t) = \varkappa Bt^{\alpha-1}n(\tau,0)\exp(-Bt^{\alpha}/\alpha) \qquad (2.25)$$

which means that $I(\tau,t)$ should depend on time like

$$I(\tau,t) = at^{\alpha-1}\exp(-Bt^{\alpha}/\alpha) \qquad (2.26)$$

It is the case, cf. Figs. 2.14 and 2.15, in which the solid lines correspond to the values calculated according to eqn. (2.26).

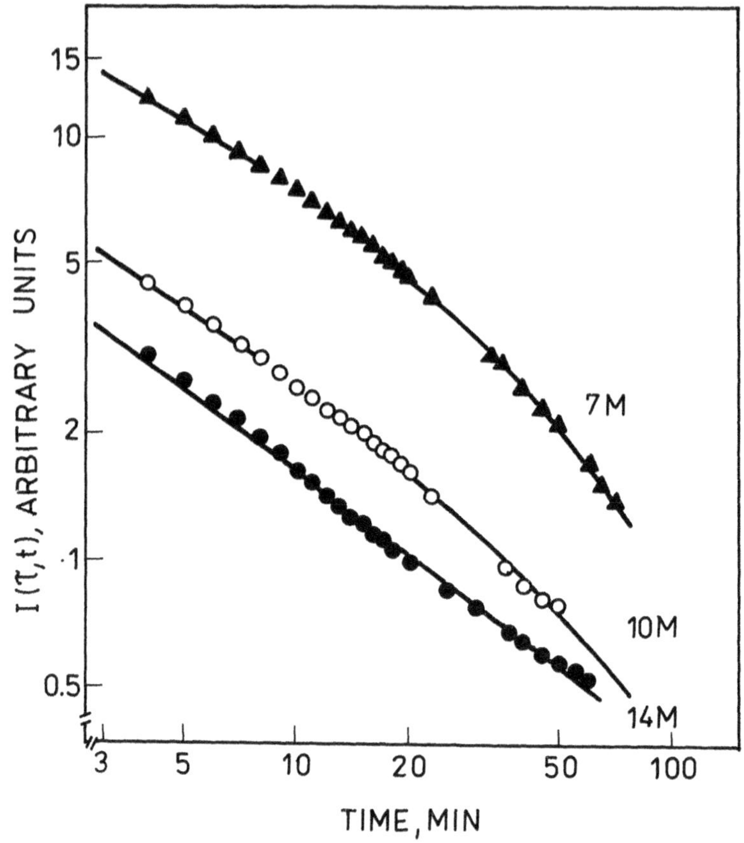

Fig. 2.14. Isothermal luminescence of LiCl glasses at irradiation temperature of 77 K. Solid lines calculated according to eqn. (2.26) with numerical values of α and B given in Table 2.1

Table 2.1. Numerical values of rate parameters for ITL and decay of RTL

Concn., M	Temp., K	ITL			RTL decay	
		a*	α	$10^2 \cdot B$**	α	$10^2 \cdot B$
7	77	24.5	0.64	5.67	0.40	5.35
7	87	37.5	0.67	6.41	-	-
10	77	10.3	0.55	5.52	0.21	5.24
14	77	7.15	0.41	2.05	-	-

* in units $min^{1-\alpha}$

** in units $min^{-\alpha}$

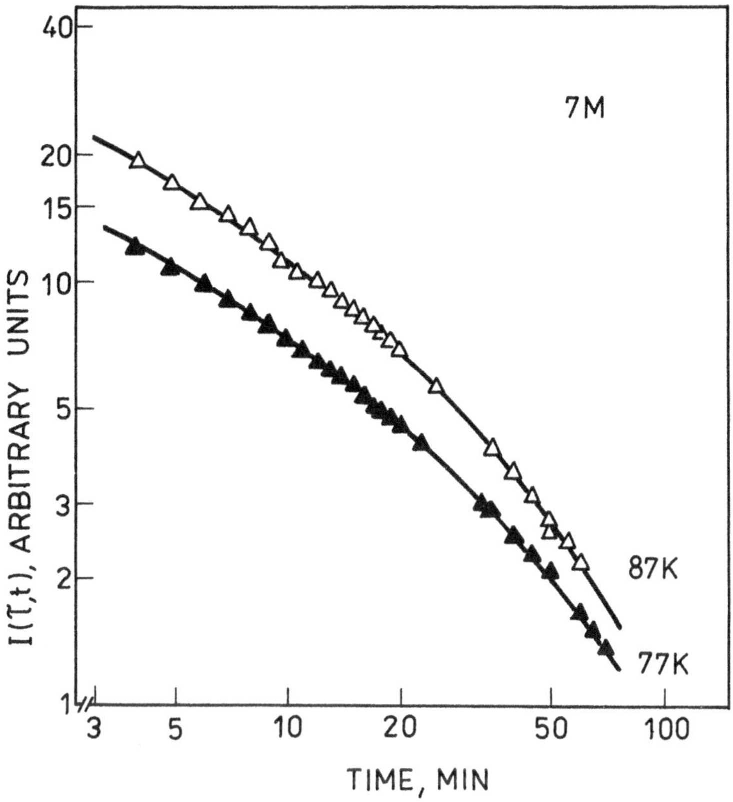

Fig. 2.15. Isothermal luminescence at 77 and 87 K of 7 M LiCl glass irradiated at 77 K. Solid lines calculated according to eqn. (2.26) with numerical values of α and B given in Table 2.1

During thermoluminescence measurements the temperature of the sample is increased linearly with time

$$T(t) = T(0) + \beta t \text{ , or} \qquad dT = \beta dt \tag{2.27}$$

Integration of eqn. (2.23) between the limits of initial temperature T_i and final temperature T gives the thermoluminescence intensity

$$I(\tau, t, \beta, T) = \alpha n(\tau, t)\, k(T,t) \exp\left[- \int_{T_i}^{T} \frac{k(T,t)}{\beta} dT\right] \tag{2.28}$$

i.e. the thermoluminescence glow curve. The area under the glow peak, denoted as A(t)

$$A(t) = \varkappa n(\tau, t) \exp\left[-\int_{T_i}^{T_f} \frac{k(T,t)}{\beta} dT\right] \qquad (2.29)$$

is proportional to the concentration of trapped electrons at time t, $n(\tau,t)$, which changes with time according to eqn. (2.24). Therefore, the area below the peak of the thermoluminescence glow curve should decrease with the time of sample storage at a constant temperature as

$$A(t)/A(0) = \exp(-Bt^{\alpha}/\alpha). \qquad (2.30)$$

Again, it is the case, cf. Fig. 2.16, presenting the decay of the first RTL peak at about 100 K (Plonka and Kowalski 1977) during the storage of γ-irradiated 7 and 10 M LiCl glasses at the irradiation temperature of 77 K.

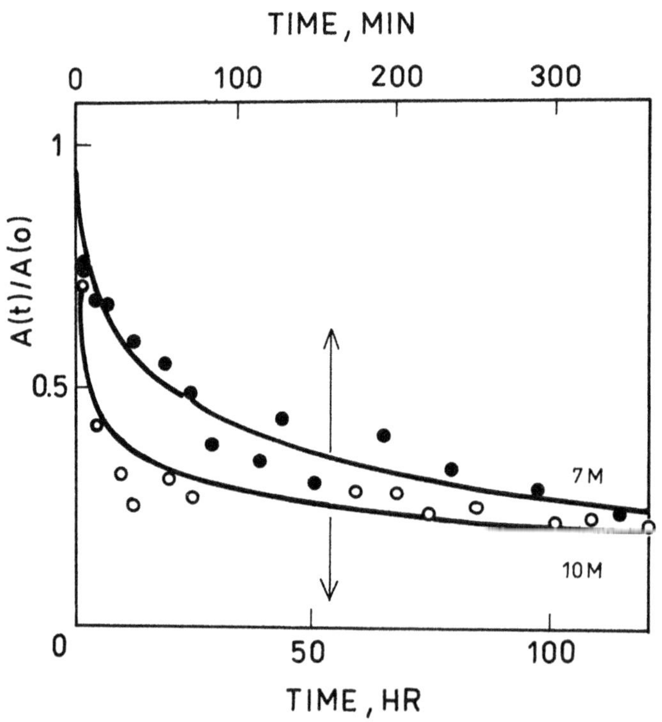

Fig. 2.16. Decay of first RTL peak for 7 and 10 M LiCl glasses stored at an irradiation temperature of 77 K. Solid lines calculated according to eqn. (2.30) with numerical values of α and B given in Table 2.1

At this point, before further analysis, it is worth to mention two other kinetic treatments of isothermoluminescence commonly used in radiation chemistry.

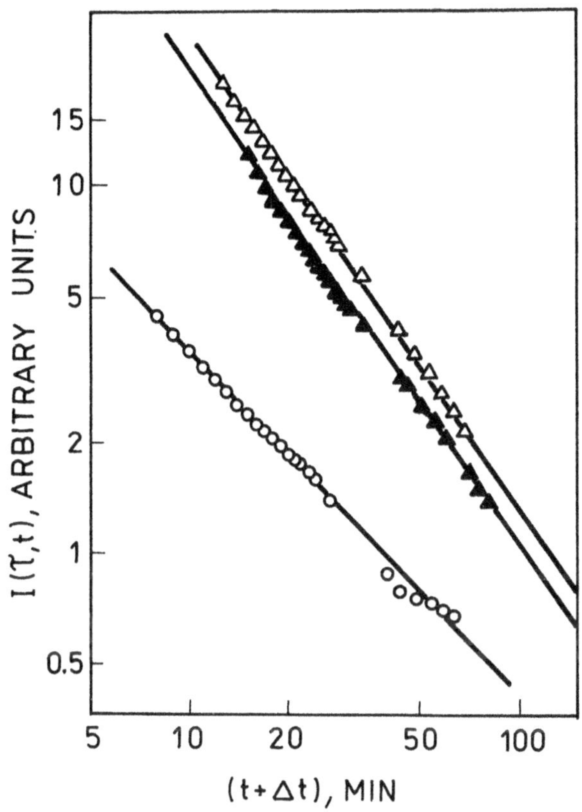

Fig. 2.17. Data from Fig. 2.14 for 10 M LiCl glass (○) and from Fig. 2.15 for 7 M LiCl glass at 77 (▲) and 87 K (Δ) in coordinate system ln $I(\tau,t)$ vs. $\ln(t+\Delta t)$

The first, due to Debye and Edwards (1952), consists of fitting the experimental data by

$$I(\tau,t) = at^{-m} \qquad (2.31)$$

where a and m are constants. Relation (2.31) yields the straight line in the coordinate system ln $I(\tau,t)$ vs. ln t, and is inconsistent with the experimental data, cf. Figs. 2.14 and 2.15. To make them consistent one is forced (Hamill 1976) to use the coordinate system ln $I(\tau,t)$ vs. ln $(t+\Delta t)$, cf. Fig. 2.17, but the effective time $(t+\Delta t)$ is completely meaningless as there is no reason to use different values of Δt for experiments performed under similar conditions.

The second of commonly used kinetic treatments, due to Bagdasar'yan et al. (1967), based on the use of relation

$$I(\tau,t)/I(\tau,0) = 1 + kt \qquad (2.32)$$

is valid only, cf. Fig. 2.18, for a very limited range of luminescence decay.

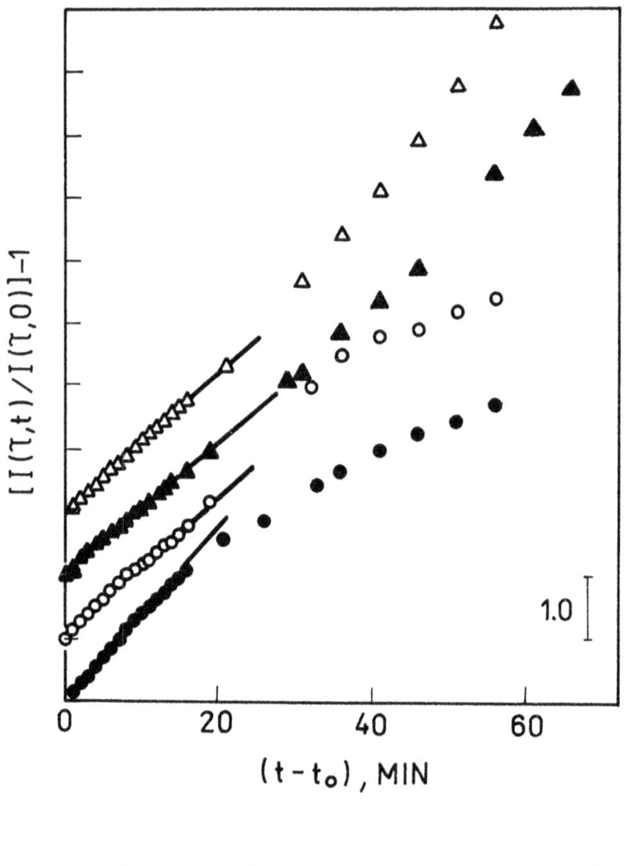

Fig. 2.18. Data from Figs. 2.14 and 2.15 plotted according to eqn. (2.32) To avoid point overlapping, curves are shifted vertically

The distributions of electron reactivities, eqn. (1.6), deduced from ITL and RTL measurements for 10 M LiCl glass are given in Fig. 2.19. It is clearly seen that ITL originates only from part of the most reactive species giving rise to RTL. This is also true for 7 M glass. With an increase of LiCl concentration the ITL intensity decreases and the reactivity distribution, as may be inferred from numerical values of α, becomes wider. It is not surprising that as the depth of shallow cationic traps decreases with the increase of salt concentration above about 7 M, the Li^+ traps are expected to lose their bonding properties

(Bartczak et al. 1981). Reports (Klassen et al. 1980) on the decrease of infrared electrons with increase of concentration are pertinent at this point.

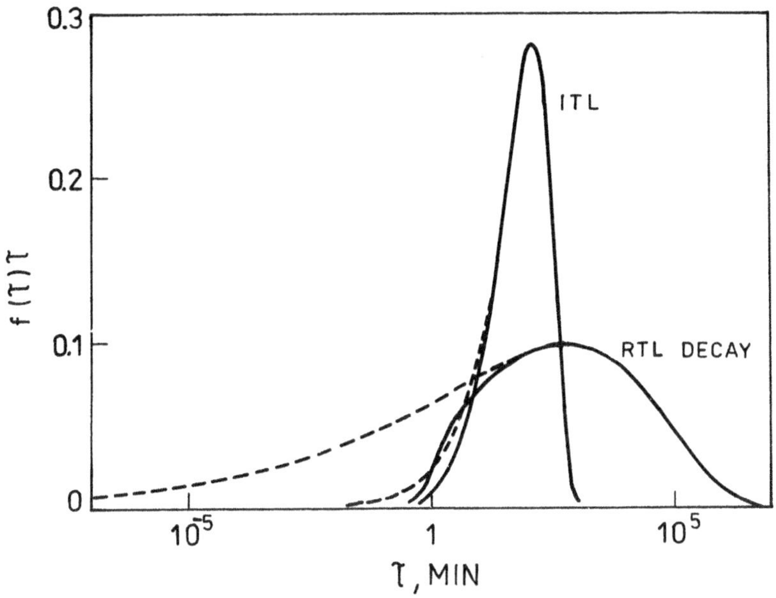

Fig. 2.19. Distributions of electron reactivities deduced from RTL decay and ITL of 10 M LiCl glass: initial distribution $f(\tau,0)\tau$ (– – –) and distribution recalculated according to eqn. (2.7) for t = 1 min $f(\tau,t)\tau$ (———)

The numerical values of α for ITL of 7 M LiCl glass at 77 and 87 K are close enough to allow for at least a rough estimation of activation energy. The value 0.2 kcal/mole coincides with that for hydrogen atom decay in aqueous glasses in the same temperature region (Plonka et al. 1983) and we do hope that this coincidence is not casual.

Thus advanced kinetic treatment of radioluminescence kinetics enabled us (Wypych et al. 1985) to check the fundamental difference in electron trapping predicted by the theoretical model of Bartczak et al. (1981) for the frozen solutions of $MgCl_2$ and $CaCl_2$. In this model, based on the lattice theory of concentrated solutions two ionic structures are considered, cation and anion vacancies, as the possible source of the attractive potential. It appears that due to the repulsive effects of the nearest neighbouring anions the cation site forms a very shallow electron trap or even has no bonding electron level. For 2.5 - 4.5 M solutions of $MgCl_2$ and $CaCl_2$, which form upon freezing glasses of nearly identical properties (Angel and Sare 1970), the potential trap depth, on average, is smaller for Ca^{2+} ions than for Mg^{2+} ions. Addition of the repulsive potentials of the neighbouring anions give shallow but still bounding traps in the case of Mg^{2+} whereas weaker Ca^{2+} ion potential is covered by anion repulsion. Analysis of the experimental results, cf. Table 2.2, supports the above prediction fairly well. The distributions of electron reactivities calculated according to the normalized saddle point approximations with (a/B) taken as the normalization factor, for numerical values of rate parameters collected in Table 2.2 are depicted in Figs. 2.20 - 2.23.

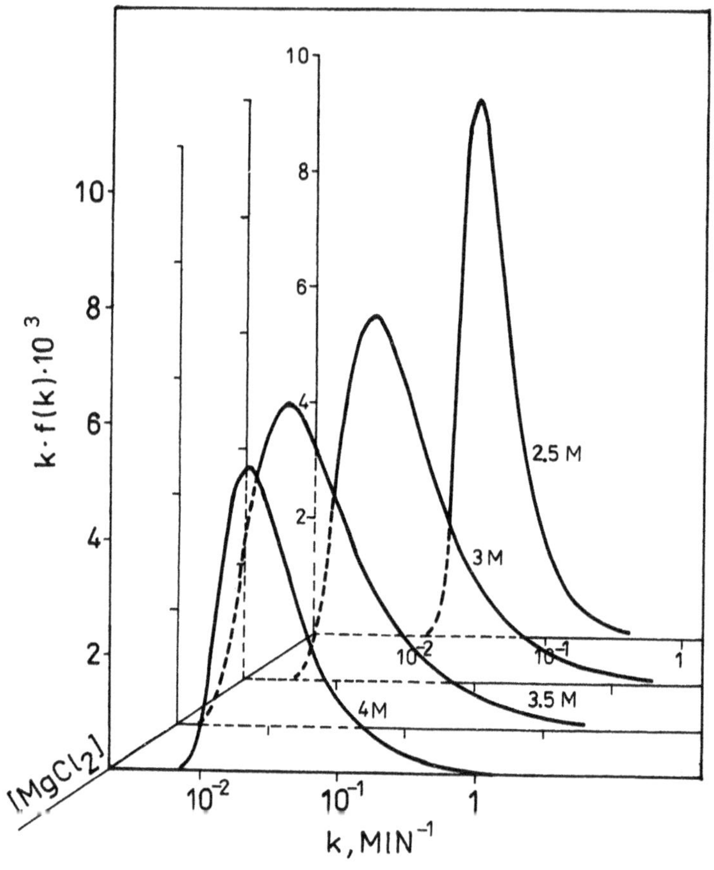

Fig. 2.20. Effect of solute concentration on reactivity distribution of electron involved in ITL of $MgCl_2$ solutions

In full agreement with the discussed model there are many more electrons giving rise to ITL in $MgCl_2$ than $CaCl_2$ solutions, cf. Figs. 2.20 and 2.21. An increase of solute concentration decreases their population markedly, as predicted for the decrease of the average cation-anion distance, shifting the reactivity distribution of the remaining part towards lower rate constants. It may indicate some contribution to ITL

Table 2.2. Numerical values of rate parameters for ITL of $MgCl_2$ and $CaCl_2$ solutions

Solute/concn. M		$NaNO_3$ concn. mM	a $min^{1-\alpha}$	$10^2 \cdot B$ $min^{-\alpha}$	α
$MgCl_2$	2.5	none	762	7.47	0.79
$MgCl_2$	3.0	none	875	6.81	0.67
$MgCl_2$	3.5	none	787	6.08	0.64
$MgCl_2$	4.0	none	635	7.14	0.71
$MgCl_2$	4.0	1.0	152	5.97	0.71
$MgCl_2$	4.0	5.0	42.6	3.67	0.71
$MgCl_2$	4.0	10	6.1	3.41	0.62
$CaCl_2$	3.0	none	234	4.93	0.66
$CaCl_2$	3.5	none	48.2	2.05	0.65
$CaCl_2$	4.0	none	33.4	3.93	0.51
$CaCl_2$	4.5	none	16.8	4.80	0.40
$CaCl_2$	4.0	1.0	6.0	2.95	0.55
$CaCl_2$	4.0	5.0	0.84	1.65	0.48

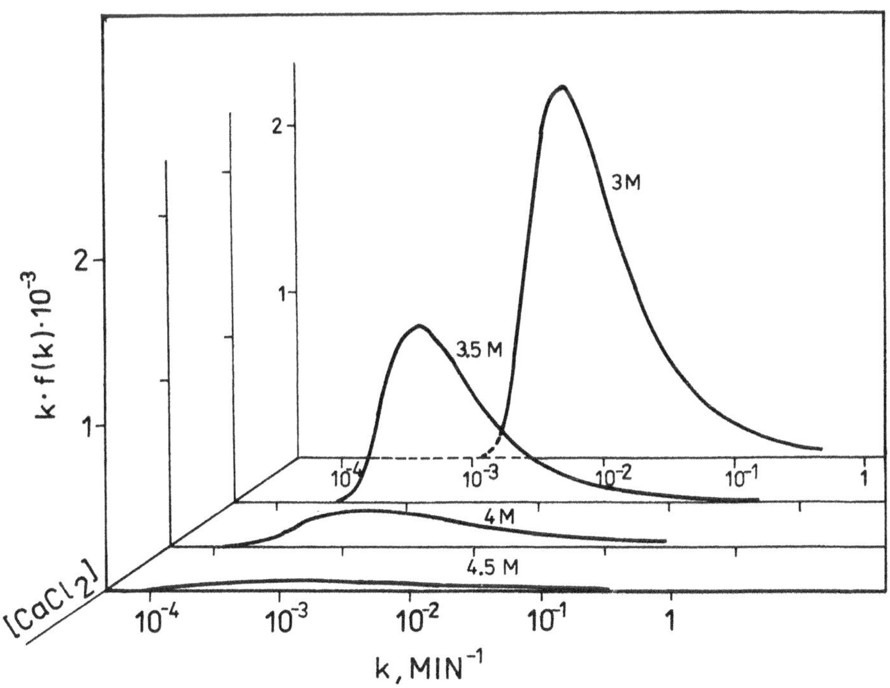

Fig. 2.21. Effect of solute concentration on reactivity distribution of electrons involved in ITL of $CaCl_2$ solutions

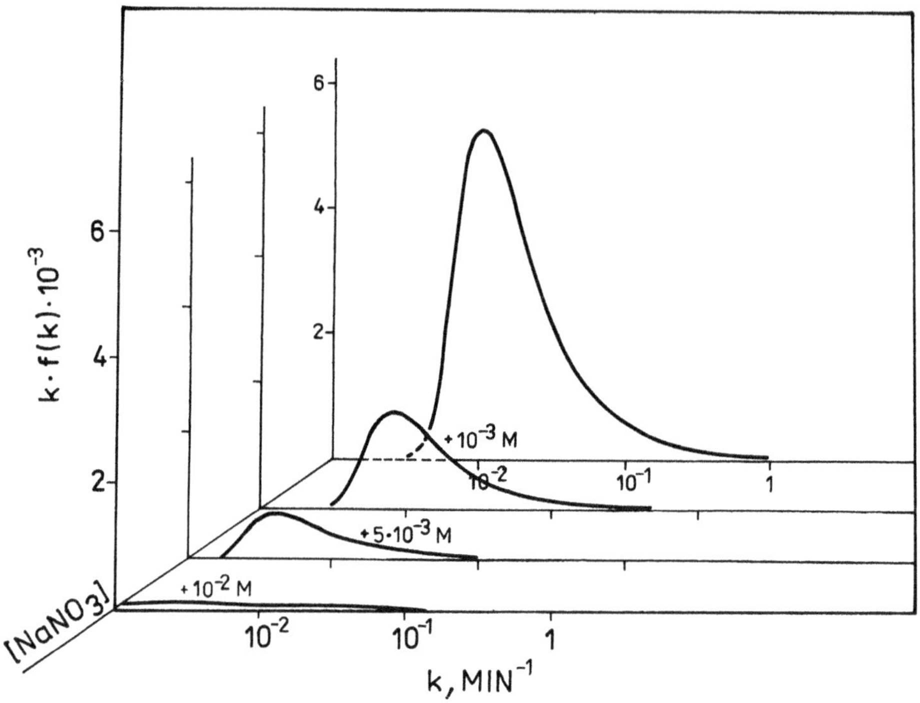

Fig. 2.22. Effect of scavenger concentration on reactivity distribution of electron involved in ITL of $MgCl_2$ solutions

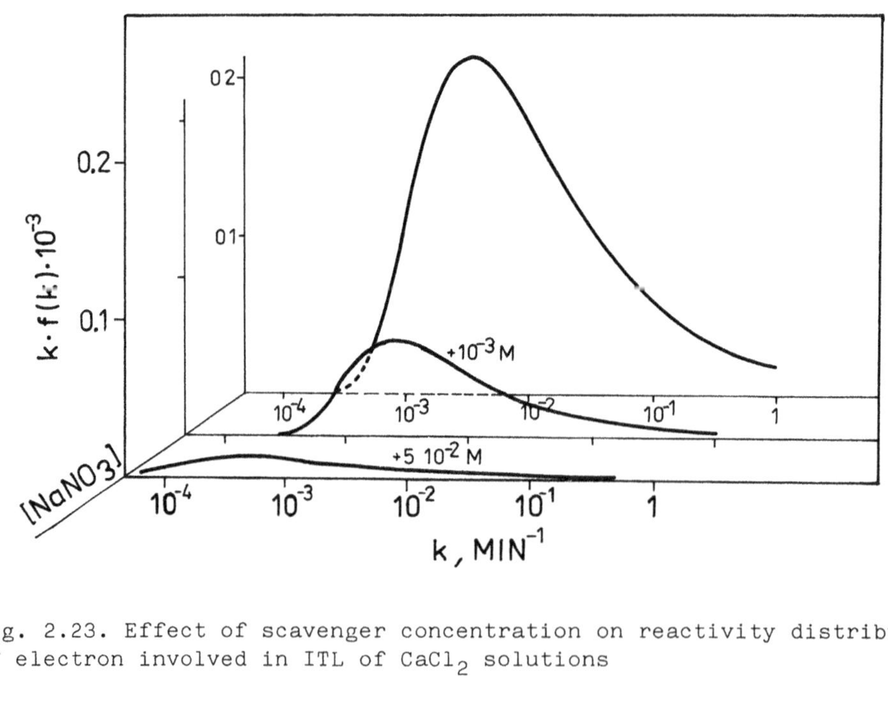

Fig. 2.23. Effect of scavenger concentration on reactivity distribution of electron involved in ITL of $CaCl_2$ solutions

of electrons in deep traps, i.e. anion vacancies as the addition of an electron scavenger has the same effect on reactivity distribution in both kinds of matrices, cf. Figs. 2.22 and 2.23. To make this point clear more complete data are needed than those published (Gillis and Walker 1976; Nguen and Walker 1977; May and Walker 1978) on the total amount of electrons trapped in $CaCl_2$ and $MgCl_2$ solutions.

HYDROGEN ATOMS

Extensive ESR studies on trapped hydrogen atoms formed in acidic ices upon γ-irradiation at low temperatures revealed that the reactivity of these atoms at a given temperature decreases with time (Sprague and Schulte-Frohlinde 1973; Kroh and Plonka 1975) and depends upon the temperature at which they were formed (Kroh and Plonka 1975). The nature of the reaction partner is secondary importance (Plonka et al. 1978), cf. Fig. 2.24 illustrating the decay of trapped hydrogen atoms in reactions with various alcohols.

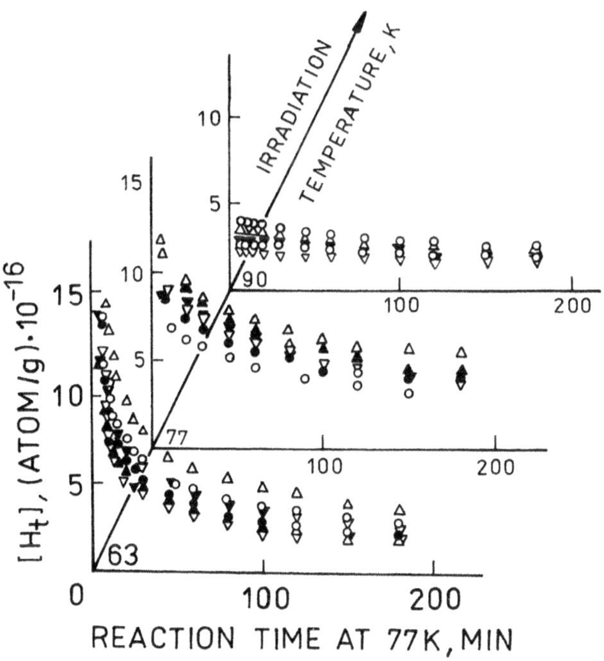

Fig. 2.24. Decay of trapped hydrogen atoms at 77 K in 6 M sulfuric acid glasses γ-irradiated at 63, 77, and 90 K. Additives: 1.0 M of methyl (○), ethyl (△), propyl (▽), iso-propyl (●), butyl (▲), sec-butyl alcohol (▼)

It is seen that all alcohols behave similarly in sulfuric acid glasses although there are marked differences in the rate constants of their reaction with hydrogen atoms in aqueous solutions, namely $1.6 \cdot 10^6$ for methyl, $2.6 \cdot 10^7$ ethyl, $2.5 \cdot 10^7$ propyl, $6.5 \cdot 10^7$ iso-propyl, $3.8 \cdot 10^7$ butyl, $1.3 \cdot 10^8\ M^{-1}s^{-1}$ for sec-butyl alcohol at 15° C (Neta et al. 1971).

To account for the decrease of reactivity with time at a given temperature one is forced to use the time-dependent rate constant in the kinetic equations for the irreversible decay of trapped hydrogen atoms due to recombination or reaction with additives. After many attempts, the adequate description of decay patterns was endeavoured by Plonka et al. (1979) with the use of the time-dependent rate constant of the form (2.2)

Kinetics of Trap-Limited Decay

The decay of trapped hydrogen atoms at 77 and 87 K in 6 M sulfuric acid glasses γ-irradiated at 77 K are presented in Fig. 2.25 (Plonka et al. 1979).

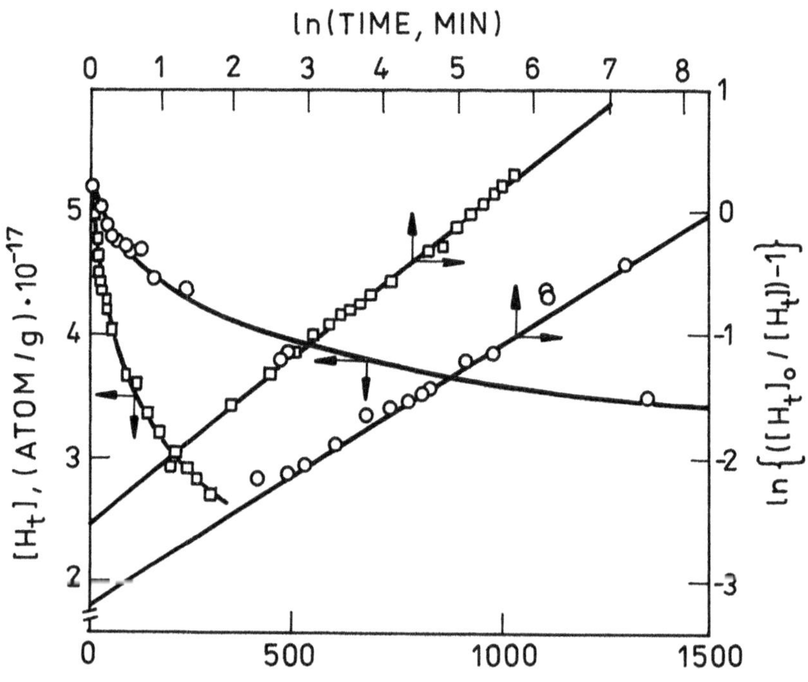

Fig. 2.25. Decay of trapped hydrogen atoms at 77 (O) and 87 K (□) in 6 M sulfuric acid glass γ-irradiated at 77 K

This decay is due almost exclusively to the recombination

$$H_t + H_t \longrightarrow H_2 \qquad (2.33)$$

which was described by the second-order equal concentration kinetic equation with time-dependent rate constant given by eqn. (2.2), i.e. by

$$(1/[H_t]) - (1/[H_t]_o) = Bt^{\alpha}/\alpha \ , \qquad (2.34)$$

which yields a straight line in the coordinate system $\ln\{([H_t]_o/[H_t])-1\}$ vs. ln t. It is evident that the experimental data fit perfectly to relation (2.34).

In the presence of 2-propanol the trapped hydrogen atoms decay due to reactions (2.33) and (2.35)

$$H_t + HC(CH_3)_2OH \longrightarrow H_2 + C(CH_3)_2OH \ , \qquad (2.35)$$

and the proper kinetic equation was assumed (Plonka et al. 1979) to be

$$-d[H_t]/dt = k(t)[H_t]^2 + k'(t)[H_t][HC(CH_3)_2OH] \qquad (2.36)$$

Taking k(t) and k'(t) in the form given by relation (2.2) and denoting by $A = B'[HC(CH_3)_2OH]$ one obtains the relation

$$-d[H_t]/dt = B[H_t]^2 t^{\alpha-1} + A[H_t]t^{\alpha-1} \ , \qquad (2.37)$$

valid for all cases when $[HC(CH_3)_2OH] \gg [H_t]$. The use of one numerical value of α is proved below experimentally.

Integration of (2.37) leads to

$$\ln\{(1/[H_t] + B/A)/(1/[H_t]_o + B/A)\} = At^{\alpha}/\alpha \qquad (2.38)$$

Relation (2.38) simplifies substantially when the contribution of reaction (2.33) in the decay is negligible in comparison with reaction (2.35). For $B/A \longrightarrow 0$ one has from (2.38)

$$\ln([H_t]_o/[H_t]) = At^{\alpha}/\alpha \ . \qquad (2.39)$$

Relation (2.39) was assumed to be proper for the description of the decay of trapped hydrogen atoms in the presence of 1 M of 2-propanol. Relation (2.39) yields a straight line in the coordinate system $\ln\ln([H_t]_o/[H_t])$ vs. ln t.

Again a perfect fit of the experimental data to relation (2.39) is evident, cf. Fig. 2.26. Furthermore, the use of one numerical value of α in eqn. (2.37) is justified. The contribution of reaction (2.33) in this case, estimated from $B[H_t]_o/(A + B[H_t]_o)$, is not greater than 3 per cent at 77 K and 1 per cent at 87 K.

The constancy of the numerical value of α for reactions at a given temperature enables the use of the relation (2.38) to describe the decay of trapped hydrogen atoms in the presence of smaller amounts of 2-propanol, i.e. in the cases when both reactions (2.33) and (2.35) contribute to the decay but $[HC(CH_3)_2OH] \gg [H_t]$ holds. The experimental data fit well to relation (2.38), cf. Fig. 2.27. The contribution of

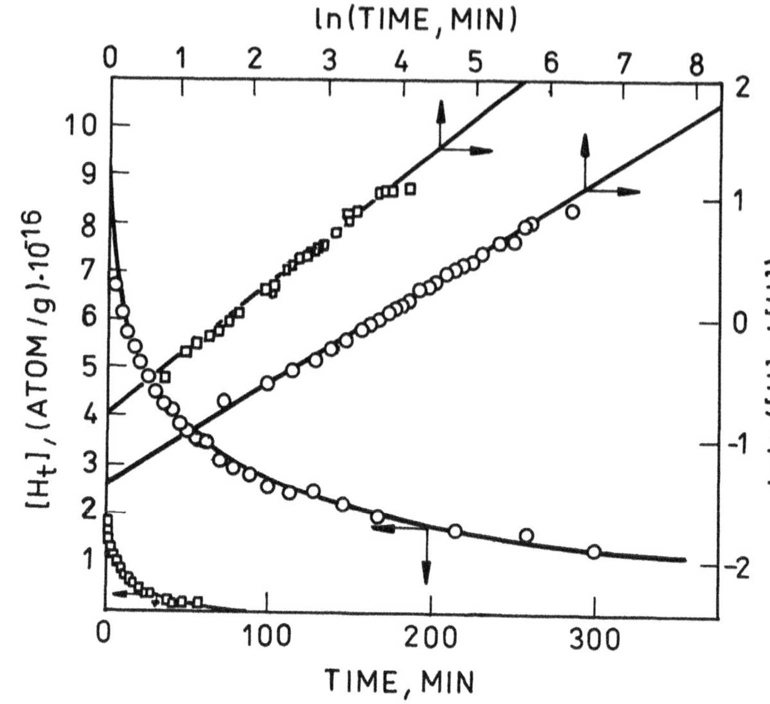

Fig. 2.26. Decay of trapped hydrogen atoms at 77 (○) and 87 K (□) in 6 M sulfuric acid glass containing 1 M of 2-propanol γ-irradiated at 77 K

hydrogen atom recombination to the decay observed in the presence of 0.1 M of 2-propanol is equal to about 30 per cent at 77 K and 22 per cent at 87 K as estimated from the ratios of $B[H_t]_o/(A + B[H_t]_o)$.

Table 2.3. Numerical values of rate parameters for trapped hydrogen atom decay in 6 M sulfuric acid glasses with and without 2-propanol added

Additive	T,K	$[H_t]_o$ [a]	α	A [b]	B [c]
none	77	5.8	0.39		2.78
none	87	6.3	0.49		6.07
1.0 M 2-Pr	77	1.1	0.38	0.094	
1.0 M 2-Pr	87	0.34	0.49	0.217	
0.1 M 2-Pr	77	4.8	0.38	0.030	
0.1 M 2-Pr	87	3.6	0.49	0.080	

[a] $(spin/g)\cdot 10^{17}$; [b] $min^{-\alpha}$; [c] $(g/spin\ min)^{-\alpha})\cdot 10^{20}$

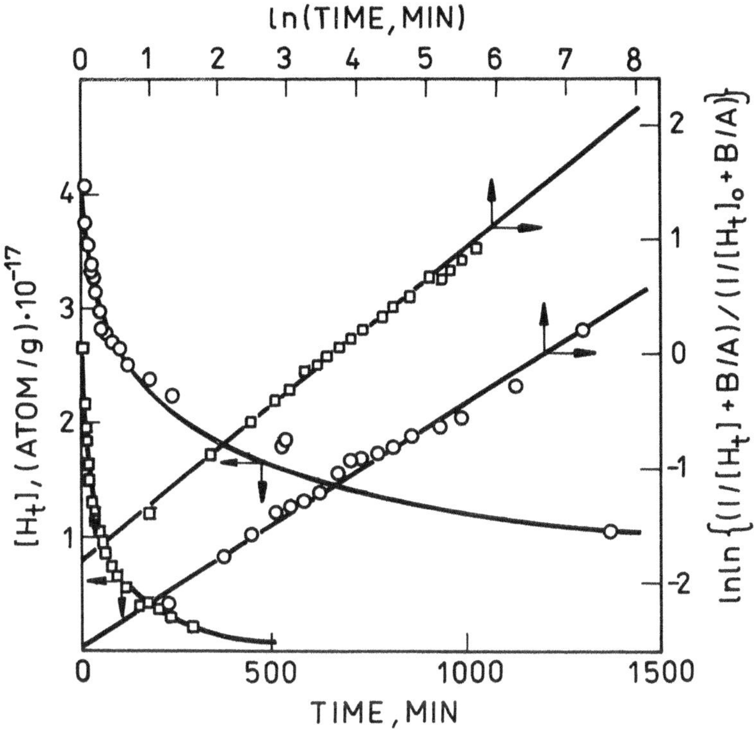

Fig. 2.27. Decay of trapped hydrogen atoms at 77 (O) and 87 K (□) in 6 M sulfuric acid glass containing 0.1 M of 2-propanol γ-irradiated at 77 K

Looking at the numerical values of α and B collected in Table 2.3 it is worth mentioning two points. The first point is that the numerical values of α, 0.38 - 0.39 at 77 K and 0.49 at 87 K, are not too different from 0.5 to discredite the previously reported curve fittong results in the coordinate systems $[H_t]_o/[H_t]$ or $\ln([H_t]/[H_t]_o)$ vs. $t^{0.5}$ (Vacek 1977) or vs. $(-at-2bt^{0.5})$. However, the classical diffusion kinetics leading to these relations is not adequate and the fit is reasonable only for decays in short, limitted intervals. The second point is that the temperature dependence of the decay patterns is contained in those of A or B and α which are related to the detail of environmental interaction on trapped hydrogen atoms and structure of amorphous materials. We will return to this point below, showing at first the effect of irradiation temperature on the rate of decay.

Effect of Irradiation Temperature on Decay Rate

Fig. 2.28 presents the isothermal decay at 77 K of trapped hydrogen atoms in γ-irradiated at 63, 77 and 87 K 6 M sulfuric acid glasses containing 1 M of 2-propanol (Plonka et al. 1982). At this concentration of 2-propanol the reaction of trapped hydrogen atoms with 2-propanol

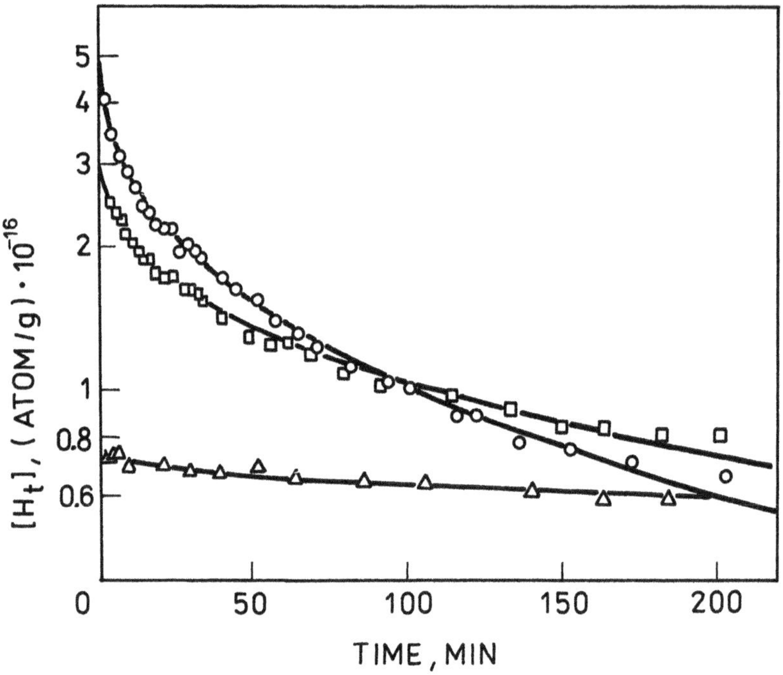

Fig. 2.28. Decay of trapped hydrogen atoms at 77 K in 6 M sulfuric acid glass containing 1 M of 2-propanol γ-irradiated at 63 (○), 77 (□), and 87 K (△). The solid lines drawn through the experimental points are calculated from relation (2.39) with numerical values of α and A given in Table 2.4

Table 2.4. Numerical values of rate parameters for trapped hydrogen atom decay at 77 K in γ-irradiated 6 M sulfuric acid glasses containing 1 M of 2-propanol

T_{irr}	$[H_t]_o$ [a]	α	A [b]	τ_o [c]
63	5.9	0.38	0.116	22.7
77	3.4	0.38	0.075	62.3
87	0.77	0.38	0.013	7200

[a] (spin/g)$\cdot 10^{-16}$; [b] $min^{-\alpha}$; [c] min

was shown to dominate over the recombination and the pseudo first-order decay with time-dependent rate constant is observed. All experimental data fit perfectly to the relation (2.39), i.e. to the pseudo first-order kinetics equation with time-dependent rate constant (2.2).

Here it is worth noting that the particular form (2.2) of the rate constant was known to be consistent with the long-tail jump time distribution function for a random walk in disordered material, see above. It is also consistent, however, with the empirical expression for the relaxation function

$$\Phi(t) = \exp\left[-(t/\tau_o)^{\alpha}\right], \quad 0 < \alpha \leqslant 1 , \qquad (2.40)$$

which has been found to give an excellent fit to a wide variety of relaxation processes in vitreous systems, cf. Introduction. Indeed, from relation (2.40) one can obtain the time-dependent rate constant in the form

$$k(t) = -\left[d\Phi(t)/dt\right]/\Phi(t) = \alpha\tau_o^{-\alpha}t^{\alpha-1} \qquad (2.41)$$

which is identical with (2.2) for

$$B = \alpha\tau_o^{-\alpha} \qquad (2.42)$$

The relaxation time τ_o in (2.40) depends in part x on the absolute temperature T and in part (1-x) on the "structural" temperature T_F

$$\tau_o = C \exp\left[\frac{x\Delta h^*}{RT} + \frac{(1-x)\Delta h^*}{RT_F}\right] \qquad (2.43)$$

where Δh^* denotes the activation enthalpy and the "structural" temperature T_F is defined as the temperature at which the instantaneously arrested structure at T corresponds to the equilibrium structure.

Making use of relation (2.42), it is possible to present eqn. (2.39) in the form

$$[H_t] = [H_t]_o \exp\left[-(t/\tau_o)^{\alpha}\right], \qquad (2.44)$$

more suitable for further analysis. According to (2.44) all experimental data for pseudo first-order decay should give one straight line in the coordinate system $\ln([H]/[H]_o)$ vs. the reduced time $(t/\tau_o)^{\alpha}$. It is the case, cf. Fig. 2.29. The numerical values of τ_o, let us call it effective lifetime, cf. Appendix B, are seen, cf. Table 2.4, to depend markedly on the temperature at which the trapped hydrogen atoms were formed. They do also depend on the temperature at which the decay is followed. For glasses irradiated at 77 K the effective lifetime at 87 K, according to the data presented in Table 2.3, is equal to 5.27 min.

Relation (2.43) predicts that the relaxation time τ_o depends in part x on the absolute temperature and in part (1-x) on the "structural" temperature T_F. However, T_F could not be identified with the irradiation temperature T_{irr}. To find the relationship between T_F and T_{irr} we recall that with the use of T_F the departure of the instantaneous value of the monitored property from its equilibrium value is measured in the temperature units (DeBolt et al. 1976). The equilibrium concentration of

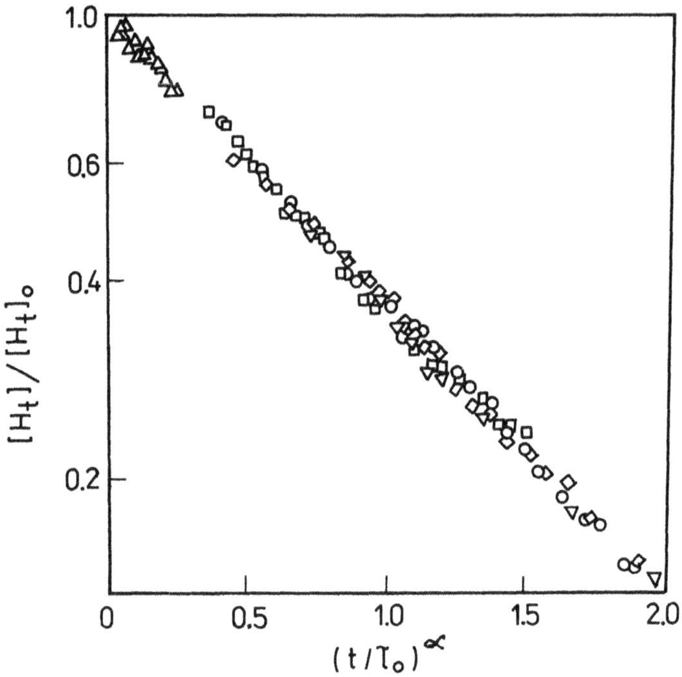

Fig. 2.29. Experimental data from Fig. 2.28 (○), (□), (△), and Fig. 2.26 (◇), (▽) plotted according to the relation (2.24)

trapped hydrogen atoms at any temperature is zero and thus the departure from the equilibrium concentration is just equal to the observed concentration. To express it in temperature units we may simply say that the "structural" temperature is proportional to the trapped hydrogen atom concentration and to look for the dependence of $[H_t]_o$ on T_{irr}.

It was found, cf. Fig. 2.30, that within the experimantal errors of the data

$$[H_t]_o = a(T_f - T_{irr}) \ , \tag{2.45}$$

and $\ln\tau_o$ is linear in $1/(T_f - T_{irr})$, where T_f = 92.5 K is the threshold temperature for freeing the hydrogen atoms in 6 M sulfuric acid glasses. The physical sense of T_f is illustrated in Fig. 2.31 which presents the results of thermal annealing of 6 M sulfuric acid glass γ-irradiated at 77 K. At T_f, some 50 degrees below T_g, the very rapid decay od trapped hydrogen atoms starts and the first peak begins.

Returning to the main point, one can use formula (2.43) with

$$T_F = T_f - T_{irr} \tag{2.46}$$

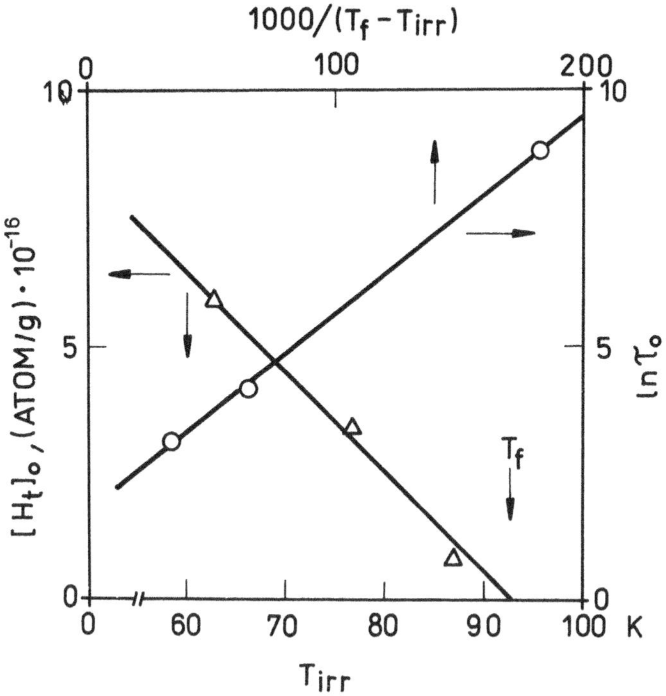

Fig. 2.30. Dependence of $[H_t]_o$ on T_{irr} and $\ln\tau_o$ on $1/(T_f - T_{irr})$

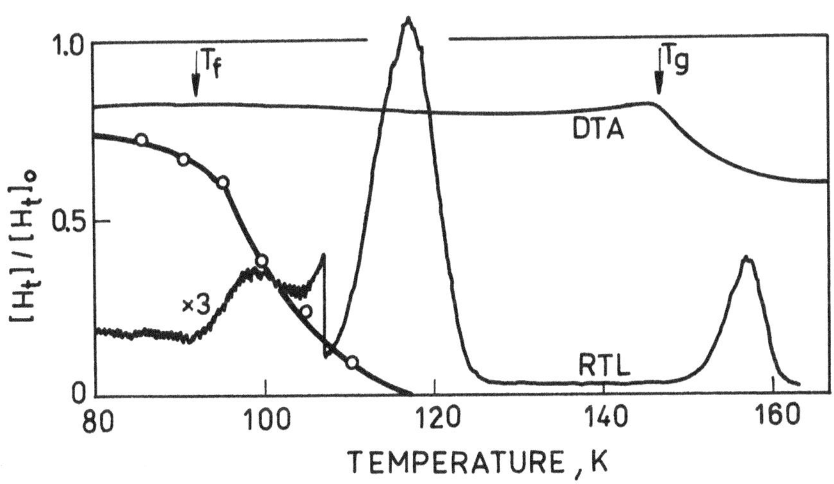

Fig. 2.31. Comparison of thermal annealing with DTA and RTL curves for 6 M sulfuric acid glass γ-irradiated at 77 K

to describe quantitatively the effect of T_{irr} on τ_o at the given temperature. For the discussed experimental data, cf. Fig. 2.30, one has

$$\tau_o = \exp(1.8 + \frac{38.5}{92.5 - T_{irr}}) , \qquad (2.47)$$

Making use of the data from Table 2.3 for the decay of trapped hydrogen atoms at 77 and 87 K in the same system γ-irradiated at 77 K one can proceed further and introduce the temperature at which the reaction is followed into (2.47). The result is

$$\tau_o = C \exp(\frac{1350}{T} + \frac{38.5}{92.5 - T_{irr}}) \qquad (2.48)$$

and that completes the formula for dependence of τ_o on T and $T_{irr} > T_f$ = = 92.5 K. The preexponential factor, depending mainly on scavenger concentration, is equal to $1.43 \cdot 10^{-7}$ min for the present system, and the activation enthalpy Δh^*, cf. (2.43), is equal to about 2.7 kcal/mole.

Mechanism of Decay

Returning now to the temperature dependence of the kinetic parameters A or B and α , discussed above, it becomes desirable to compare the decay at various irradiation temperatures starting each time from the same concentration of trapped hydrogen atoms formed at comparable radiation yields. This was achieved with the use of X-band ESR spectrometer coupled with van de Graaff accelerator (Plonka 1983). To avoid trouble with spurious signals from hydrogen atoms trapped in quartz flow Dewar, the single spin-flip satellite line was monitored, cf. Fig. 2.32, some 7.5 G apart from the main high field ESR line due to trapped hydrogen atoms in 6 M sulfuric acid glass. In that way it was possible to follow the accumulation of trapped hydrogen atoms under irradiation with electron beam and subsequent decay, cf. the example shown in Fig. 2.33, in the temperature range 88 - 113 K. The experimental data were found to fit perfectly, cf. Fig. 2.34, to relation (2.34) with α equal to 0.39 ± 0.01. The slopes of the lines presented in Fig. 2.34, equal to $[H_t]_o B/\alpha$, are shown in Fig. 2.35 in Arrhenius coordinate system. Because of the constancy of α and $[H_t]_o$ this figure illustrates the temperature dependence of B. Into this figure there is included the experimental result, cf. Fig. 2.25, for 6 M sulfuric acid glass γ-irradiated at 77 K (α = 0.39).

The necessary ESR background for discussion of the presented kinetic data was provided by investigations of Shiraishi et al. (1976) on hydrogen atoms formed in pure and acidic ices under continuous electron irradiation. In the presence of acidic solutes they were able to observe the hydrogen atoms from about 100 K, the lowest attainable temperature in their system, to the vicinity of the melting point, while in pure ice hydrogen atoms were detectable only below 140 and above 160 K. In acidic ice below 130 K the spectral features of hydrogen atoms, like hyperfine splitting and line width, were about the same as those at 77 K. The intensity of hydrogen atom signal decrease with temperature increase. The narrowing of line width beginning at about 130 K was accompanied by a marked decrease of signal intensity; at about 140 K the signal was hardly discernible. Much narrower and clearer signals reappear above

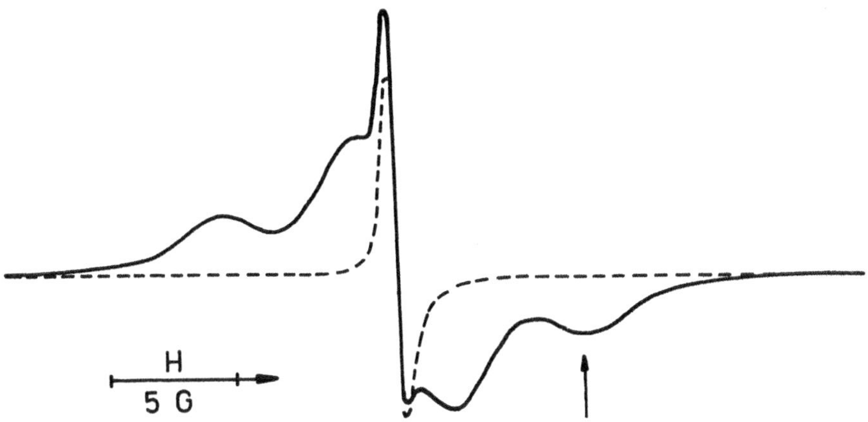

Fig. 2.32. The high-field ESR line of hydrogen atoms formed in 6 M sulfuric acid glass (solid line) and quartz (dashed line under electron irradiation at 113 K. The arrow points to the single spin-flip satellite line monitored continuously in the discussed experiments

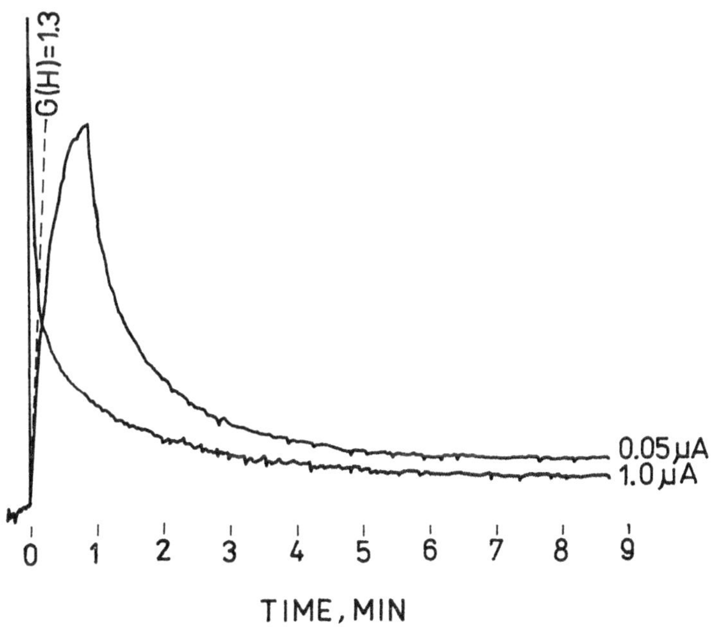

Fig. 2.33. Accumulation of hydrogen atoms in 6 M sulfuric acig glass under irradiation with electron beam and subsequent decay as observed by continuous monitoring of single spin-flip satellite line at 113 K

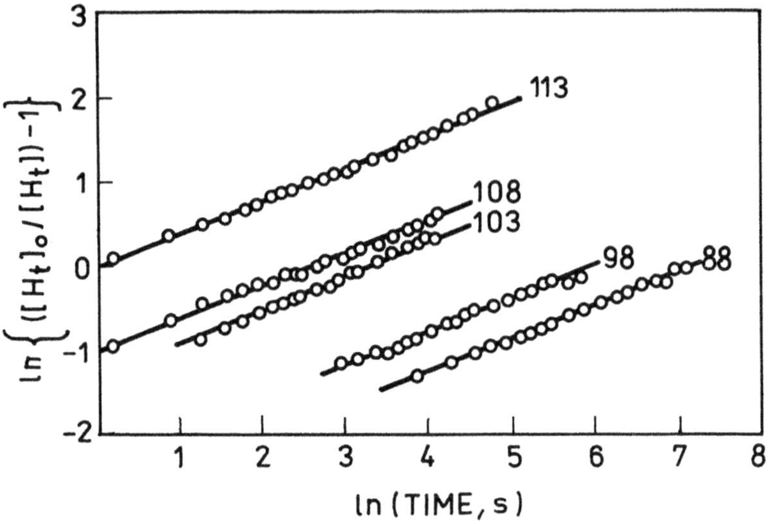

Fig. 2.34. Fit of experimental data to eqn. (2.34)

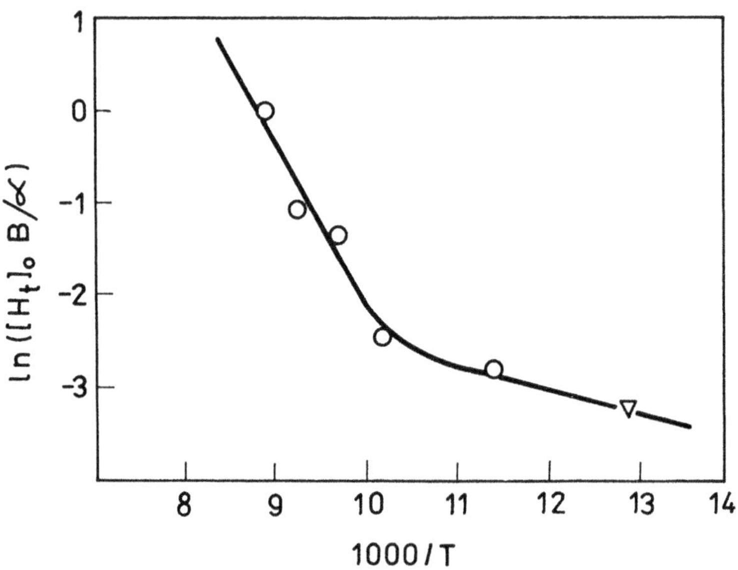

Fig. 2.35. Non-Arrhenius behaviour of the constant B

160 K but they were oppositely polarized, i.e. the low-field line was in emission while the high-field line was in absorption.

Two possible interpretations of line width narrowing were considered:

tumbling motion of water molecules surrounding trapped hydrogen atoms and translational diffusion of hydrogen atoms. Comparison with dielectric relaxation and nuclear spin relaxation data, indicating very slow molecular reorientation, favoured the second possibility. Thus below 130 K the presence of acidic solutes suppresses effectively the translational diffusion of hydrogen atoms and we are dealing with so called trapped hydrogen atoms. Furthermore, in this temperature range there is no doubt that the ESR signal is strictly proportional to hydrogen concentration.

In the above discussion it was stressed that molecular reorientation coupled with a certain specific defect may be very fast and that such motion would probably lead to an escape of the hydrogen atom trapped within the cage of the defect. This is the most relevant point for our discussion which let us start with the original model (Sprague and Schulte-Frohlinde 1973) of trap-limited hydrogen atom decay. In this model the question of transport mechanism was not addressed and the kinetic behaviour was explained as follows. Immediately after irradiation almost all hydrogen atoms are located in shallow traps which are present in large numbers. They move from trap to trap through the matrix until they encounter some reactive species with which they react and thus disappear or until they encounter a relatively deep trap from which they are not able to escape quickly. The overall process is a competition for the hydrogen atoms between reactive species and deep traps.

The question of transport mechanism was addressed when the lack of isotope effect at 63 - 90 K was evidenced, cf. Fig. 2.36 (Kroh and Plonka 1975). Hydrogen and deuterium atoms were assumed to be localized in similar traps in the given matrix. Because of large differences in masses the difference in probabilities for escaping from intact trap by traversing any potential barrier, either classically or quantum mechanically, should be significant while the probabilities for destruction of a trap containing either hydrogen or deuterium atom should be similar. Thus it was assumed that the trapped species are released due to trap destruction. The probability of trap destruction due to molecular rearrangement should be lower in a more relaxed system containing fewer frozen - in configuration which are unstable at the given temperature. It is true, as we have seen above, cf. Fig. 2.28. The rate of trapped hydrogen atom decay at 77 K is greater in matrix irradiated at 63 K and lower in matrix irradiated at 87 K than in the sample irradiated at 77 K. Because of this the kinetic behaviour was interpreted rather as a redistribution of hydrogen atoms into more and more stable traps, than into deeper and deeper traps, proceeding in competition with irreversible decay in reactions with other reactive species present in the matrix.

Both pictures suggested that the environment of trapped hydrogen atoms changes with time at a given temperature and indeed significant alteration of hydrogen atom microscopic environment was observed during storage at 77 K. According to recent determinations (Plonka and Bogus 1980) immediately after γ-irradiation there is 2.7 near neighbour protons at the effective distance of 2.15 Å from trapped hydrogen atoms. After 250 hrs of storage at 77 K the effective number of protons increases

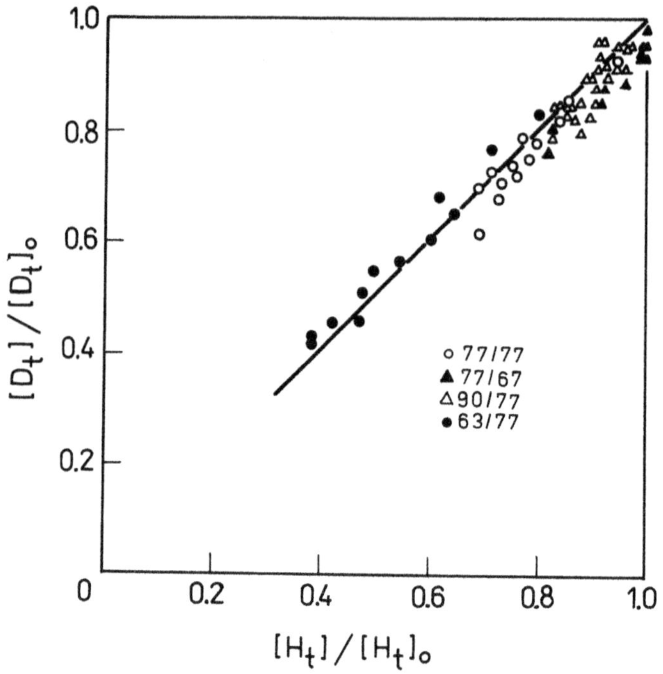

Fig. 2.36. Recombination of trapped hydrogen and deuterium atoms in 6M sulfuric acid glasses

to 4.1 and the effective distance increases to 2.37 Å. These changes do result in increased interaction of trapped hydrogen atoms with environment as evidenced (Plonka 1980) by the ESR parameters which change in time of storage at 77 K achieving the apparent equlibrium values, A = 1411.1 MHz and g = 2.0020 250 hrs after γ-irradiation. Immediately after γ-irradiation the hyperfine splitting constant A is significantly higher, A = 1412.5 - 1412.9. The observed increase of A shift from the atom value, 1420.4, provides a direct evidence of the increasing interaction with matrix.

Returning to the question of transport mechanism we are allowed to treat the hydrogen atom trap as some kind of defect in the random network structure of aqueous glass, which like amorphous solid water at 77 K, is likely to consist of hexagons of both "boat" and "chair" forms (Rice 1975). Within the trap the hydrogen atom is free to move as evidenced by angular modulation of END and EED interactions (Plonka and Kevan 1976). Translational movement, however, to the nearest cavity involves some molecular rearrangement. It may be the formation of "chair" hexagon in the trap walls, cf. the mechanism of hydrogen atom diffusion in the liquid water with apparently the same random network structure (Benderskii et al. 1980). The molecular rearrangement involved for translational motion of hydrogen atom, cf. above, seems to be very fast in pure ice even below 50 K, cf. Chapter 3. In acidic glasses the presence of solutes suppresses effectively this kind of rearrangement and translational movement of hydrogen atom is arrested. Here it is worth mentioning the model of Montroll and Bendler (1984) for a frozen dipole amidst a swarm of mobile defects, discussed in Chapter 6, as a possible modification of this mechanism. However, one immediate consequence of the above picture is that there is no qualitative distinction between mobile and trapped hydrogen atoms. The distinction is purely quantitative and so called trapped hydrogen atoms are in fact very slowly diffusing through the matrix at low temperatures. The rate limiting step is the reorientation of water molecules to form the proper structure for transition from one cavity to the other.

The temperature dependence of hydrogen atom recombination patterns at 88 - 113 K is contained in B, cf. Fig. 2.35. Its non-Arrhenius behaviour provides an additional support for the proposed mechanism of hydrogen atom recombination. At about 100 K the recombination rate increases abruptly. In this temperature range there is the first peak of thermally stimulated current (TSC) observed upon heating of water vapour condensate (Onsager et al. 1978). It is directly related to dipolar relaxation processes in the condensate and decreases markedly with the increase of condensation temperature. At the same temperature there is observed the first peak of thermally stimulated depolarization (TSD) current for polycrystalline ice (Johari and Jones 1975). It is due to the relaxation of frozen - in orientation polarization of water molecules in ice. Both these peaks have a relaxational origin analogous to thermoluminescence and first radiothermoluminescence peak, cf. Fig. 2.31, matches both. All these stress the importance of first peak in the density of vibrational states of solid amorphous water at about 70 cm^{-1} ($\Theta \sim 90$ K) (Plonka 1983).

The activation energy for the rate limiting step changes from about 0.2 kcal/mole in the low temperature region to about 3.7±0.5 kcal/mole in the high temperature region. The latter value fits perfectly to that determined by Shiraishi et al. (1976), 3.8±1.0 kcal/mole, from the correlation time for hydrogen atom diffusion in the same temperature range.

There is one further important consequence of the above presented picture of hydrogen atom movement. In some cases the rate limiting step of hydrogen atom freeing can dominate the reaction pattern. To illustrate this Fig. 2.37 shows the data on hydrogen atom recombination at 87, cf. Fig. 2.25, 100 and 102 K in 6 M sulfuric acid glass irradiated at 77 K. At 87 K the classical second-order plot is concave toward the horizontal axis and this kind of decay pattern can be adequately described by the second-order kinetics with time-dependent rate constant (2.2), cf. Fig. 2.25 presenting this data in the coordinate system $\ln([H_t]_o/[H_t] - 1)$ vs. ln t. At 100 and 102 K the classical second-order plots show concavity toward the vertical axis, apparent acceleration of recombination which cannot be fitted with the second-order kinetics with time-dependent rate constant (2.2) for numerical values of $\alpha \leq 1$. It can be fitted, however, with the first-order kinetics with the time-dependent rate constant (2.2), cf. Fig. 2.38, and the explanation is

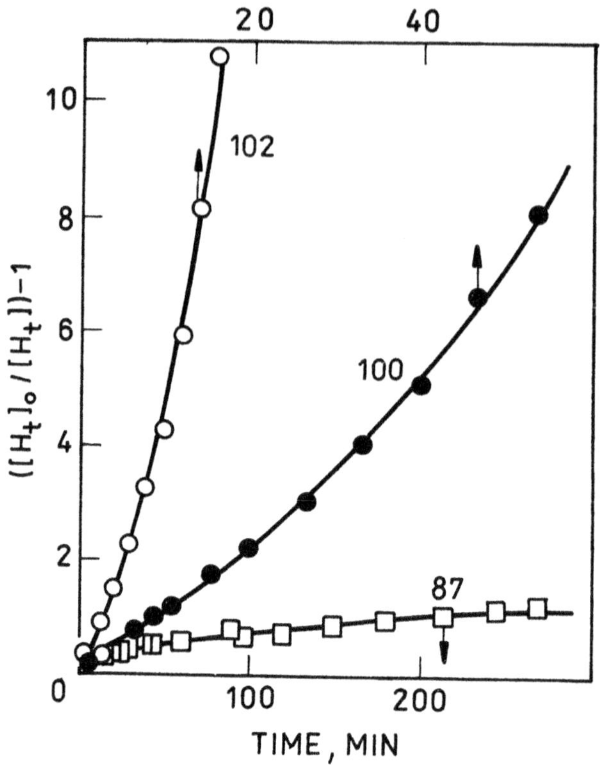

Fig. 2.37. Decay of trapped hydrogen atoms at 87 (□), 100 (●), and 102 K (○) in 6 M sulfuric acid glasses irradiated at 77 K

that the decay pattern is dominated by the first-order freeing of hydrogen atoms. Such changes of reaction order for radical recombination or disproportionation are witnessed in many systems, cf. below, when the temperature of radical generation is well below the temperature of radical freeing in a given matrix and the temperature at which the decay is observed is close to or higher than that of radical freeing. If radicals are generated at the temperature at which the decay is followed second-order decay pattern is observed, cf. Fig. 2.34.

Bearing in mind the possible change of reaction order for recombination it is worth to look at hydrogen atom reactivity in some other systems.

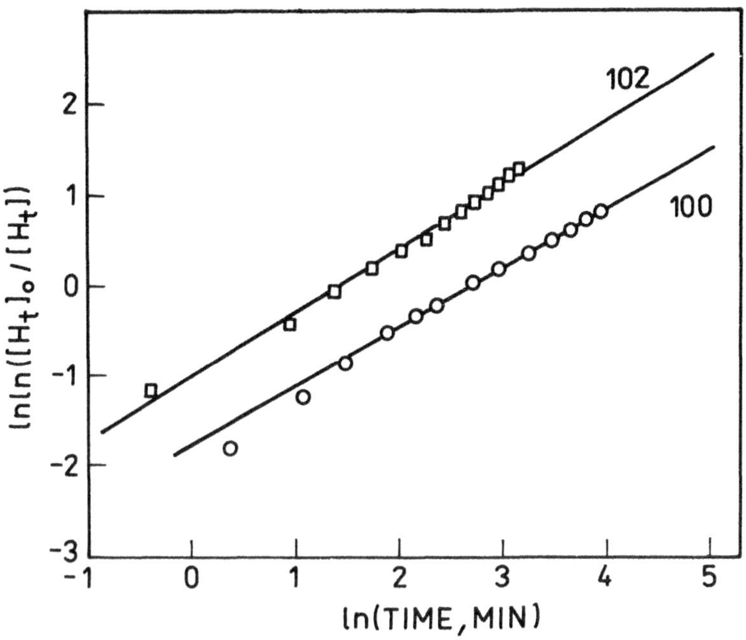

Fig. 2.38. Data from Fig. 2.37 for decay of trapped hydrogen atoms at 100 and 102 K in 6 M sulfuric acid glass irradiated at 77 K plotted in the coordinate system ln ln($[H_t]_o/[H_t]$) vs. ln t. Temperature of hydrogen atom freeing in 6 M sulfuric acid glass, cf. Fig. 2.31, is equal to 92.5 K

Effect of Matrix Dynamics on Decay Patterns

To relate the decay patterns to the dynamic properties of matrices the decay of hydrogen atoms was investigated in HCl-H_2O-$ZnCl_2$ system for the following compositions: $HCl \cdot 4\ H_2O$ (I), $H_2ZnCl_4 \cdot 12\ H_2O$ (II), $H_2ZnCl_4 \cdot 10\ H_2O$ (III), and $HZn_2Cl_5 \cdot 4\ H_2O$ (IV)(Szajdzinska-Pietek et al. 1982). Dynamics of glasses I - IV was studied previously by DTA and RTL (Plonka and Szajdzinska 1980). Two RTL maxima are characteristic for these glasses: low-temperature peak (A) located at 109 - 113 K, depending on the composition of the matrix, and high-temperature peak (B) in the vicinity of the glass transition temperature. For matrix IV the third peak (C) at 151 K was observed additionally. For all these matrices the decay of trapped hydrogen atoms at 77 K in the presence of 1 M of 2-propanol obeys the pseudo first-order kinetic equation; the value of α being equal on average to 0.53, cf. Fig. 2.39. This first-order kinetic behaviour was expected because in the presence of relatively high concentration of 2-propanol hydrogen atoms react almost exclusively with the scavenger, cf. Fig. 2.26. In the absence of 2-propanol the decay of hydrogen atoms is due to recombination and one would like to describe it by the second-order equal concentration kinetics. Such description is quite adequate for the decay at 87 K in matrices II - IV, cf. example shown in Fig. 2.40 for matrix II.

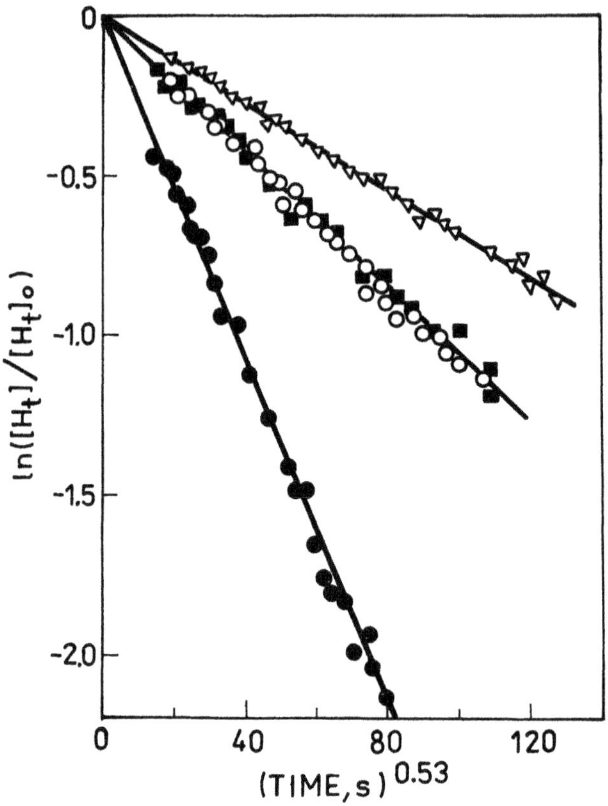

Fig. 2.39. Kinetics of hydrogen atom decay in $HCl \cdot 4\ H_2O$ (●), $H_2ZnCl_4 \cdot 12\ H_2O$ (○), $H_2ZnCl_4 \cdot 10\ H_2O$ (▽), and $HZn_2Cl_5 \cdot 4\ H_2O$ (■) at 77 K in the presence of 1 M of 2-propanol

For matrix I, however, the apparent second-order rate constant was found to increase with time and the experimental data were fitted to the first-order kinetic equation, cf. Fig. 2.41.

This difference in decay pattern seems to be due only to the difference in matrix dynamics. As suggested by Hodge and Angell (1978) in the system HCl-H_2O-$ZnCl_2$ two relaxation processes may occur below the glass transition temperature, T_g, with the intensities depending on glass composition: β process with the maximum at ca. 0.80 T_g and γ process at even lower temperatures. It may be anticipated that while the RTL peak B is connected with the main relaxation process α, i.e. with glass transition, the low-temperature RTL peaks are related to the secondary relaxations. The RTL peak A in matrix I is located at 0.82 T_g and apparently corresponds to β relaxation. Its position moves away from T_g with increase of $ZnCl_2$ content, very likely due to appearance of γ relaxations. Consistent with this suggestion is the fact that for matrix III, the RTL curve exhibits a shoulder above the maximum (Plonka

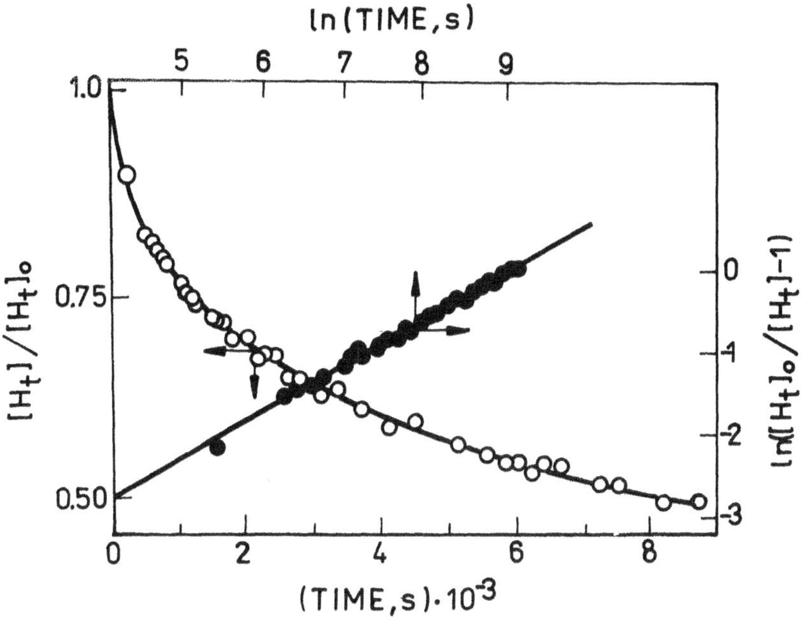

Fig. 2.40. Kinetics of hydrogen atom decay in $H_2ZnCl_4 \cdot 12\ H_2O$ at 87 K

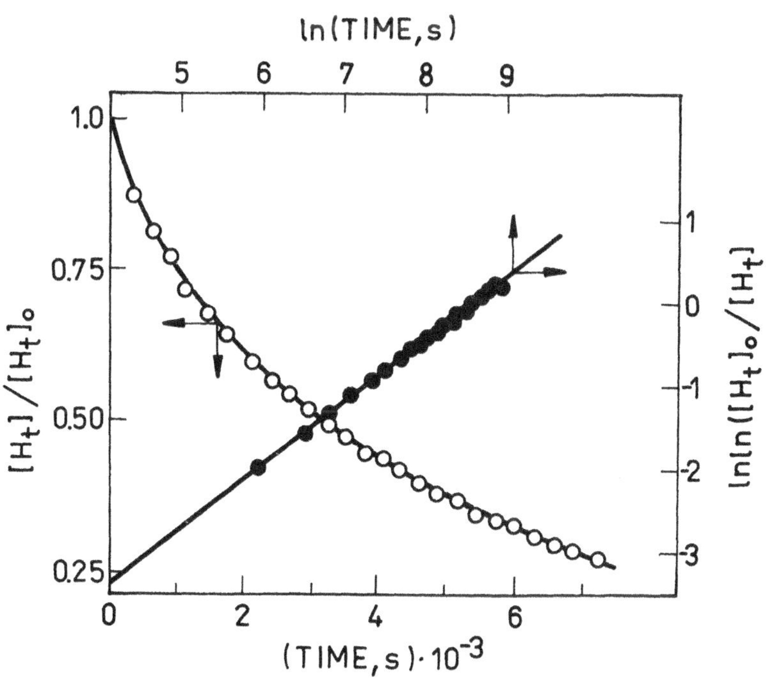

Fig. 2.41. Kinetics of hydrogen atom decay in $HCl \cdot 4\ H_2O$ at 87 K

and Szajdzinska 1980). Two RTL maxima for matrix IV at ca 0.59 T_g, peak C, and 0.81 T_g, peak A, most probably correspond to γ and β processes, respectively. Accepting these suggestions one can say that the higher the temperature of RTL peak A, the lower is the intensity of secondary relaxations at 87 K. Comparing the numerical values in Table 2.5 one finds that increase of structural relaxations is followed by an increase of the numerical value of α for the second-order decay: it rapidly approaches unity; for matrix with the highest intensity the first-order decay is observed.

Table 2.5. Numerical values of rate parameters for hydrogen atom decay in $HCl-H_2O-ZnCl_2$ matrices at 87 K; the temperature of RTL peak A are included

matrix	reaction order	α	$\tau_o \cdot 10^{-3}$, s	T_A, K
I	I	0.77	4.98	109
II	II	0.56	8.22	112
III	II	0.38	29.28	113
IV	II	0.82	9.84	110

As a further example of the effect of matrix dynamics on decay patterns let us take the data of Bhattacharya and Willard (1981). They have shown that trapped hydrogen and deuterium atoms are formed by radiolysis at 10 K of 0.5 mol% of CH_4, C_2H_6, 3-methylpentane-h_{14} (3MP-h_{14}), 3-methylpentane-d_{14} (3MP-d_{14}), C_2H_5OH or i-C_4H_{10} in Xe, of 3MP-d_{14} in Kr, and of C_2H_6 or 3MP-d_{14} in Ar. The thresholds for the onset of trapped hydrogen atom decay were at about 20 K in Ar, 30 K in Kr, and 40 K in Xe. Illustrative for the sharpness of temperature thresholds were the observations: decay of D_t in Ar-3MP-d_{14} is slow at 19 K but 75% decay in 10 min at 20.5 K; decay of D_t in Kr-3MP-d_{14} was undetectable in 5 min at 31 K but was 90% complete in 1 min at 36 K; no decay of H_t in Xe-3MP-h_{14} was detected in 2 min at 39 K but it was readily observed at 40.7 K; no decay of D_t from Xe-3MP-d_{14} was detected in 4 min at 46 K but it was 80% complete in 4 min at 48 K. While trying to fit the classical kinetic equation to the experimental data they found that all of the first-order plots are convex toward the origin, except for that for H_t in Xe-3MP-h_{14} at 48.5 K where decay is 80% complete in 1 min and apparent straight line is observed. The second-order plots show an initial concavity toward the left-axis terminated by a straight portion, except for the 40.5 K experiment for which a relatively rapid rise during the initial 4 min is followed by a straight line for the remaining 26 min. For $1.1 \cdot 10^{-2}$ M of D_t produced by 5 min X-irradiation of 3MP-d_{14} in Ar at 5 K clean second-order kinetics was observed for a fourfold decay at 20.5 K with a rate constant of 3.3 10^3 $M^{-1}s^{-1}$ for the assumed reaction $D + D \longrightarrow D_2$. In the second-order plot, $([H_t]_o/[H_t] - 1)$ vs. t, the rapid initial rise followed by a straight line observed for H_t in Xe-3MP-h_{14} at 40.7 K is indicative for second-order kinetics with time-dependent rate constant of the form (2.2). Indeed, these experimental data replotted in the coordinate system $\ln([H_t]_o/[H_t] - 1)$ vs. ln t,

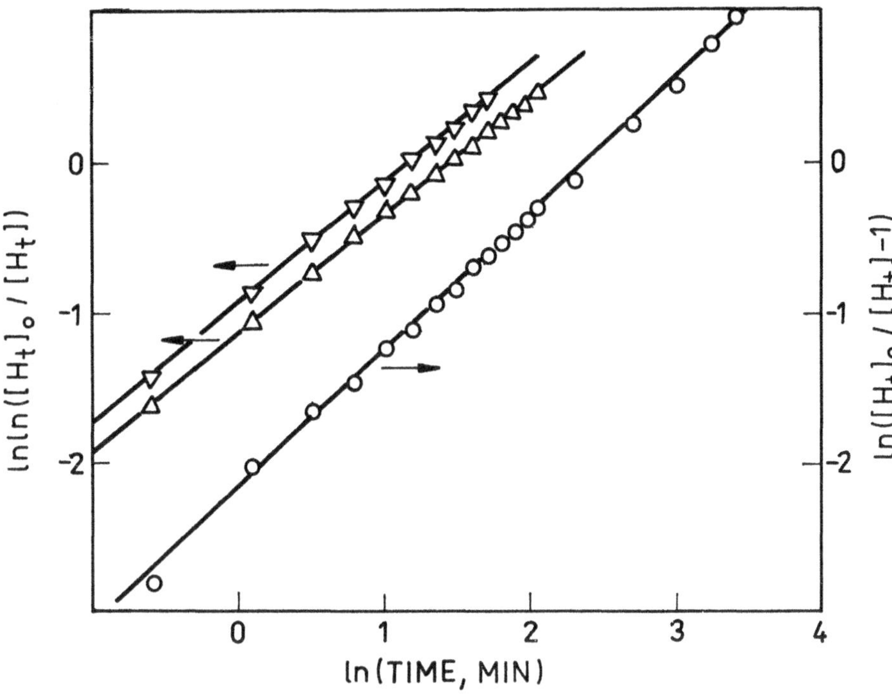

Fig. 2.42. Decay of hydrogen atoms produced by X-radiolysis of 3MP-h_{14} in Xe at 10 K: $4.9 \cdot 10^{-3}$ M of H_t at 40.7 K (O), $2.3 \cdot 10^{-3}$ M of H_t at 45.5 K (Δ), and $4.4 \cdot 10^{-3}$ M of H_t at 45.5 K (∇)

cf. Fig. 2.42, yield a straight line with the slope α equal to about 0.9. Concavity toward the left-hand axis terminated by a straight line in the second-order plot ($[H_t]_o/[H_t]$ - 1) vs. t is indicative for the first-order kinetics, cf. Fig. 2.37 and the above discussion on the change of reaction order for recombination. The experimental data for H_t in Xe-3MP-h_{14} at 45.5 K yield straight lines when plotted in the coordinate system ln ln($[H_t]_o/[H_t]$) vs. ln t, cf. Fig. 2.42, with slopes α equal to about 0.8. In contrast to trapped hydrogen and deuterium atoms the trapped 3MP-h_{14} and 3MP-d_{14} radicals in Xe are stable up to 100 K. They do decay rapidly at 100 - 110 K but the decay curves were not reported.

Miyazaki et al. (1984) have used the first-order kinetic equation with time-dependent rate constant of the form (2.2) i.e. eqn. (2.39) to analyze the decay of hydrogen atoms trapped in γ-irradiated at 77 K fused silica and single crystal of quartz. The trapped hydrogen atoms are reported to decay over several weeks at 77 K and several minutes at 100 K. Fig. 2.43 illustrates their data for hydrogen atom decay in fused silica at 90 K for two irradiation doses. Both sets of data superimpose and fit fairly well a straight line in the coordinate system

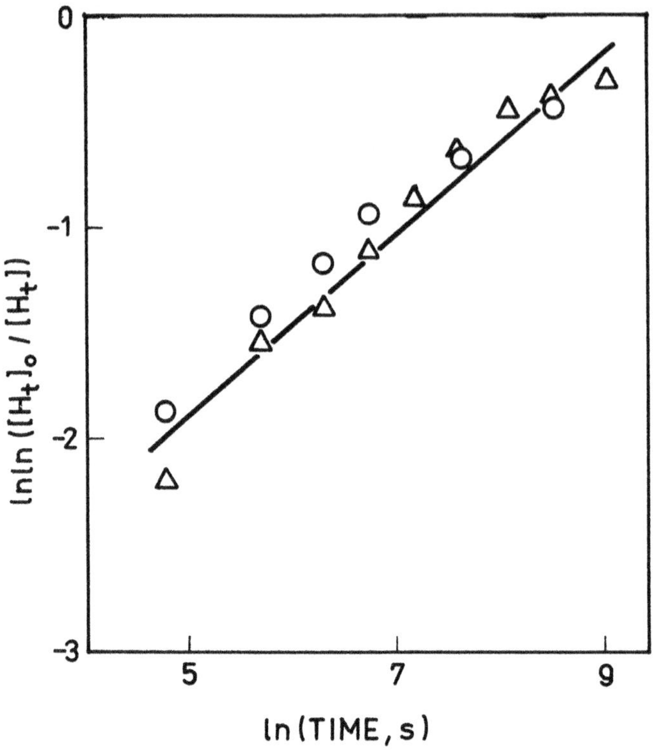

Fig. 2.43. Decay of trapped hydrogen atoms at 90 K in fused silica γ-irradiated at 77 K to $1.7 \cdot 10^5$ rads (O) and $5.1 \cdot 10^5$ rads (Δ)

$\ln \ln([H_t]_o/[H_t])$ vs. ln t.

It is interesting to note that in partially deuterated silica the decays of trapped hydrogen and deuterium atoms were found to be similar, cf. Fig. 2.44, and that the data for hydrogen atom decay in the single crystal of quartz are well fitted by eqn. (2.39).

ALKYL RADICALS

Extensive studies on the decay kinetics of radicals formed by ionizing radiation or photochemical processes in organic glasses revealed the picture surprizingly similar to that presented above for trapped hydrogen atoms. Similar kinetic patterns are observed e.g. for reactions of alkyl radical which focused the attention of several research groups since the early sixties.

Decay Patterns

It was reported by Claridge and Willard (1965) and French and Willard (1968) that CH_3 and CD_3 radicals produced from CH_3I, CH_3Br, CH_3Cl, and CD_3I by dissociative electron capture and CH_3 formed by photodissociation of CH_3I in 1 mol% solutions in 3-methylpentane (3MP) glass at 77 K do decay according to the classical first-order kinetics with the same

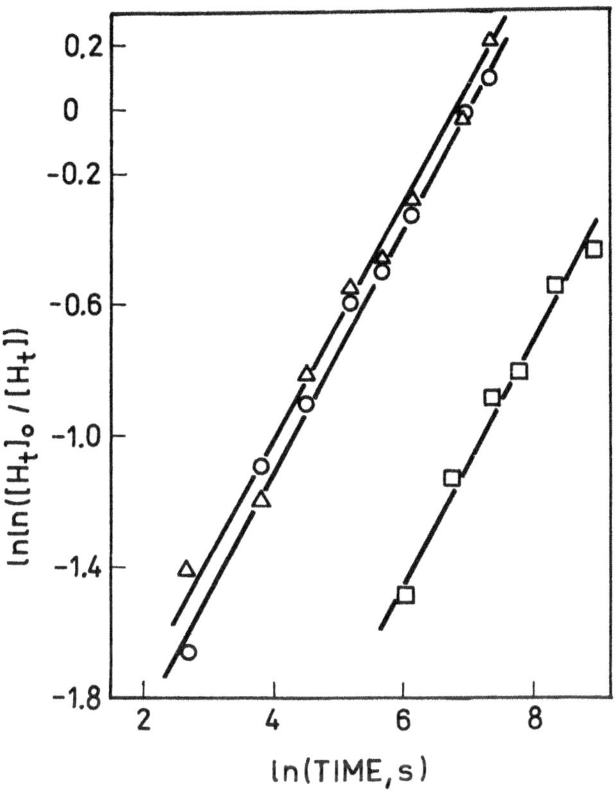

Fig. 2.44. Decay of trapped hydrogen atoms at 100 K in SiO_2 γ-irradiated at 77 K: hydrogen (O) and deuterium (Δ) atoms in partially deuterated fused silica; hydrogen atoms in single crystal of quartz (□)

16 min half-life. In other glasses, such as 2-methylpentane (2MP) or methylcyclohexane (MCHx), the half-lives were different and plots of the logarithm of the radical concentration vs. time show curvature convex toward the origin. However, plots for samples with different initial concentrations were superimpossible after normalization for initial concentration. These composite first-order decays are shown in Fig. 2.45 to be adequately described by the first-order kinetic equation with time-dependent rate constant (2.2), viz.

$$[CH_3]/[CH_3]_o = \exp(-Bt^{\alpha}/\alpha) \qquad (2.49)$$

with the rate parameters given in Table 2.6.

Furthermore, when the decays of CH_3 radicals in 3MP glass were determined below the glass transition temperature, in the range from 77 to 45 K, the plots of log CH_3 vs. time showed the curvature typical of composite first-order decay (Neiss and Willard 1975). These composite

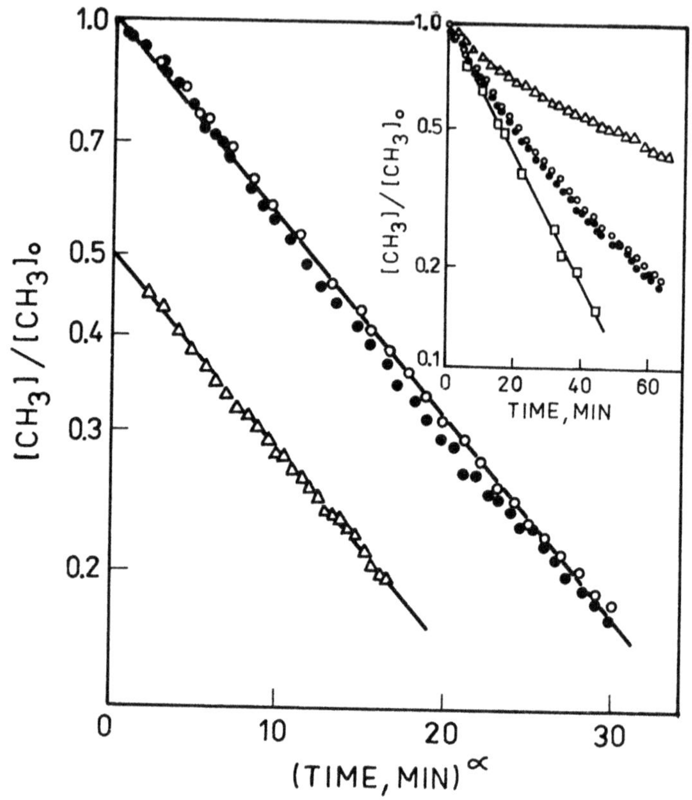

Fig. 2.45. Decay of CH_3 radicals in γ-irradiated hydrocarbon glasses containing 10^{-2} mf CH_3Cl: 2MP 2 min γdose (●), 10 min γdose (○); MCHx (Δ), 3MP (□). Insert shows original data of French and Willard (1968). Rate parameters in Table 2.6

decays are, of course, also adequately described by the first-order kinetic equation with time-dependent rate constant of the form (2.2), cf. Fig. 2.46, in full consistency with the reaction mechanism consisting in abstraction of hydrogen from 3MP (Sprague 1973; Neiss and Willard 1975).

Similar kinetics were observed for alkyl radicals produced from alkyl halides in a wide range of systems. Of particular significance for further discussion is that radical decay in a perdeuterated glassy matrix at 77 K is much slower than in the perprotiated matrix whereas the rate of decay is not influenced by deuteration of the decaying radical (Claridge and Willard 1965). This seemed to indicate that the rate-determining step in radical decay is controlled by vibrations and/or rotations of the matrix molecules, or groups of the matrix molecules, which assist the trapped radical to gain the configuration necessary for combination with the partner. The investigations of radical decay

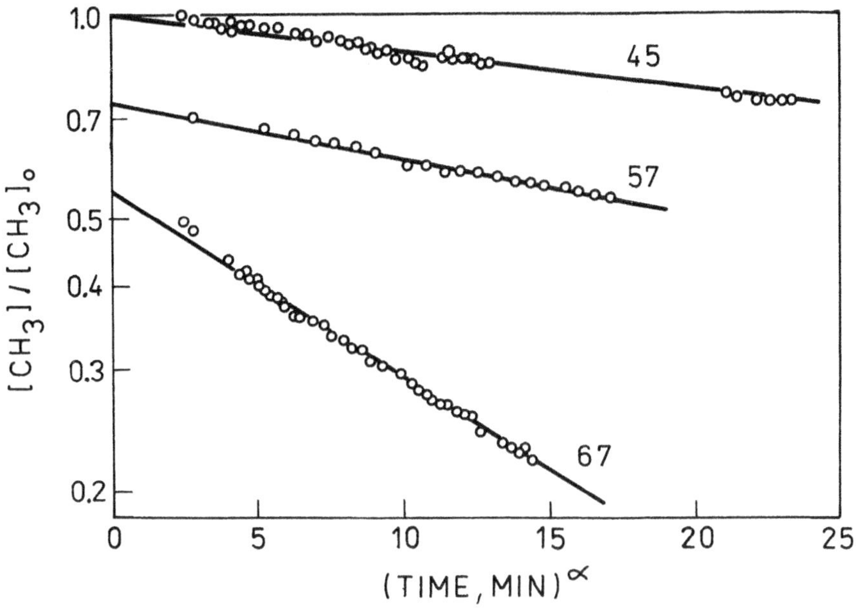

Fig. 2.46. Decay of CH_3 radicals in γ-irradiated 3MP 1% CH_3I at 67, 57, and 45 K. Rate parameters in Table 2.6

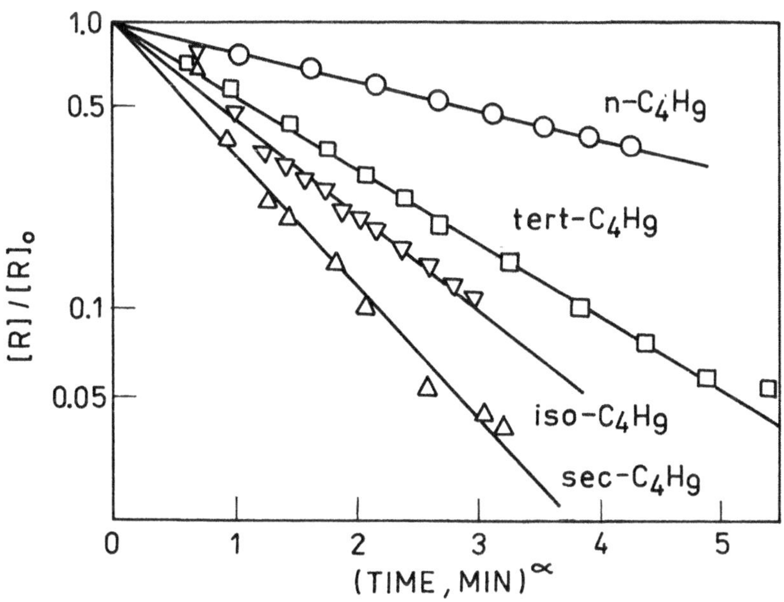

Fig.2.47. Decay of isomeric butyl radicals at 87 K formed by dissociative electron capture by 1 mol% butyl chlorides in γ-irradiated at 77 K 3MP-d_{14}. Rate parameters in Table 2.6

Table 2.6. Numerical values of rate parameters for decay of alkyl radical

Matrix	Radical	T, K	Reaction order	α	B[a]	$\bar{\tau}$ or $\tau_{1/2}$[b] min
MCHx	CH_3	77	I	0.67	0.038	97.7
2MP	CH_3	77	I	0.82	0.048	35.5
3MP	CH_3	77	I	1.00	0.044	22.9
3MP	CH_3	45	I	0.57	0.007	3517
3MP	CH_3	57	I	0.70	0.014	356.1
3MP	CH_3	67	I	0.59	0.037	168.0
3MP-d_{14}	n-C_4H_9	87	I	0.70	0.165	10.02
3MP-d_{14}	iso-C_4H_9	87	I	0.51	0.392	3.23
3MP-d_{14}	tert-C_4H_9	87	I	0.79	0.465	2.24
3MP-d_{14}	sec-C_4H_9	87	I	0.53	0.558	1.64
sec-C_4H_9Cl	sec-C_4H_9	77	II	0.79	0.0005	10414
n-C_4H_9Cl	n-C_4H_9	77	II	0.74	0.0118	267.2
iso-C_4H_9Cl	iso-C_4H_9	77	II	0.69	0.0137	294.2
3MP-d_{14}	n-C_8H_{17}	77	I	0.65	0.0566	58.49
3MP-d_{14}	n-C_4H_9	77	I	0.76	0.1070	15.53
3MP-d_{14}	C_2H_5	77	I	0.72	0.3000	4.16
n-$C_5H_{11}Cl$	C_5H_{11}	77	II	0.71	0.0069	684.8
n-$C_5H_{11}Br$	C_5H_{11}	77	II	0.66	0.0106	521.1
n-$C_5H_{11}I$	C_5H_{11}	77	II	0.62	0.0206	241.6
CD_3CD_2I	CD_3CD_2	77	II	0.37	0.0255	1374.8
CD_3CH_2I	CD_3CH_2	77	II	0.62	0.1340	11.83
CH_3CD_2I	CH_3CD_2	77	II	0.61	0.1848	7.08
CH_3CH_2I	CH_3CH_2	77	II	0.85	0.3542	2.80

[a] $min^{-\alpha}$
[b] Mean lifetime $\bar{\tau}$ for first-order reactions and first halflife for the second order reaction, cf. Appendix A

rate was extended, cf. Roy and Willard (1972), to a series of systematically chosen systems for the purpose of learning more about the relative roles of radical structure and matrix structure in determining such rates. To compare the decay rates of the four isomeric butyl radicals when produced by the same mechanism in identical matrices Roy and Willard (1972) γ-irradiated 1 mol% solutions of n-C_4H_9Cl, sec-C_4H_9Cl, iso-C_4H_9Cl, and tert-C_4H_9Cl in 3MP-d_{14} at 77 K. Fig. 2.47 illustrates their results for radical decay at 87 K in the coordinate system proper for the first-order kinetics with time-dependent rate constant (2.2). It is seen that the decay kinetics of sec-C_4H_9, iso-C_4H_9, and tert-C_4H_9 radicals is similar, that of n-C_4H_9 much slower. The respective mean lifetimes are equal to 1.64, 3.23, 2.24, and 10.02 min, cf. Table 2.6. Comparison of Fig. 2.47 with Fig. 2.48 shows that the nature of the matrix in which decay occurs affects markedly the relative decay rates of the isomeric radicals. The data in Fig. 2.48 are for the isomeric

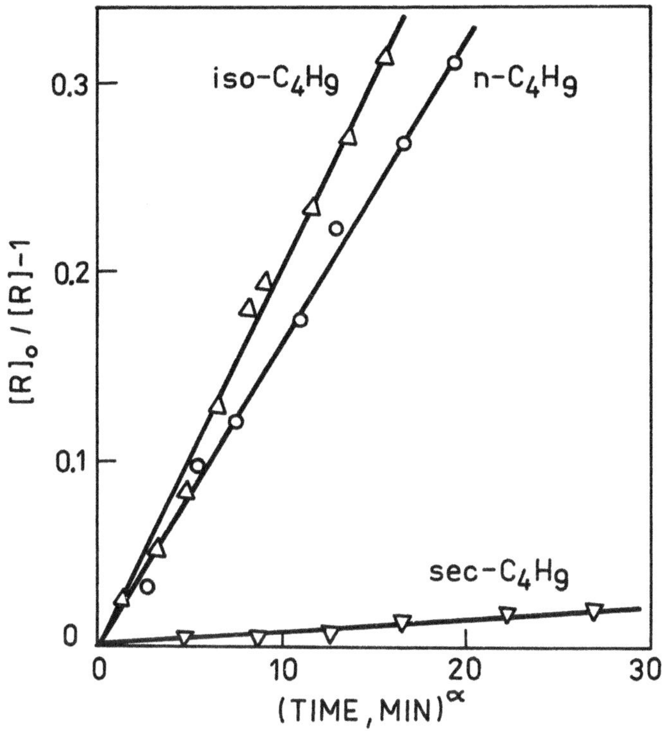

Fig. 2.48. Decay of isomeric butyl radicals at 77 K in isomeric glassy butyl chlorides γ-irradiated at 77 K. Rate parameters in Table 2.6

butyl radicals decaying at 77 K in γ-irradiated parent isomeric compounds due to recombination which is consistent with the second-order equal concentration kinetics with time-dependent rate constant (2.2), viz.

$$[R]^{-1} - [R]_0^{-1} = Bt^{\alpha}/\alpha \qquad (2.50)$$

This time the decay kinetics of n- and iso-C_4H_9 radicals is similar, recombination half-lives equal to about 267 and 294 min, that of sec-C_4H_9 much slower, recombination half-life of about 10400 min. It was not possible to obtain tert-butyl chloride in the form of glass.

When 1 mol% solutions of C_2H_5I, n-C_4H_9I, and n-$C_8H_{17}I$ in 3MP-d_{14} were γ-irradiated at 77 K and then raised to 87 K, the C_2H_5, C_4H_9, and C_8H_{17} radicals decayed at rates decreasing with increase of chain length, cf. Fig. 2.49 presenting data of Roy and Willard (1972) in the coordinate system proper for the first-order kinetics with time-dependent rate constant (2.2). The rates of decay of alkyl radicals have also been compared in their glassy chloride, bromide, and iodide matrices. For ethyl, propyl, butyl and pentyl halides decay in iodide matrices was

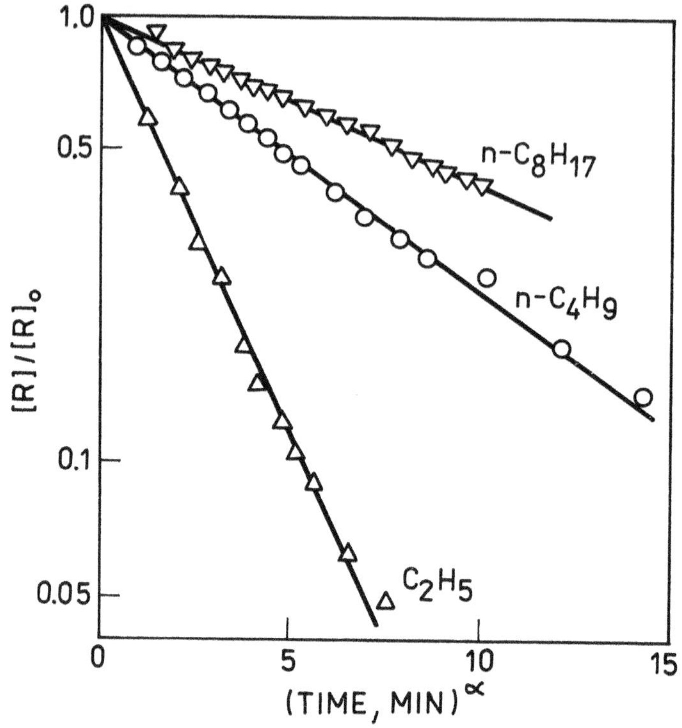

Fig. 2.49. Decay at 87 K of C_2H_5, n-C_4H_9, and n-C_8H_{17} radicals formed by dissociative electron capture by 1 mol% alkyl iodides in 3MP-d_{14} γ-irradiated at 77 K. Rate parameters in Table 2.6

faster than in the bromide matrices, which was faster than in the chloride matrices. Only for hexyl halides the order was I > Cl > Br. An example of the experimental data is shown in Fig. 2.50 for pentyl halides. Also compared were the rates of ethyl radical decay in CD_3CH_2I and CH_3CD_2I with those in C_2D_5I and C_2H_5I. The results are presented in Fig. 2.51.

All these data confirm not only the validity of the time-dependent rate constant (2.2) to describe adequately the reaction patterns but also illustrate the relative independence of decay kinetics on the nature of the potential geminate partner and the extent of deuteration of the radical. The assumption that the radical decay is determined by the rate at which radicals retrieve the configuration necessary for combination with their partners is supported by the difference seen for n-C_4H_9 radical decay in 3MP-d_{14} and the rates for the other butyl radicals, as well as the decrease of decay rate with the increasing chain length of the alkyl group for decays examined in the same matrix. The slowness of sec-C_4H_9 decay in glassy sec-C_4H_9Cl relative to the decays of n-C_4H_9 and iso-C_4H_9 in their parent matrices, as compared to the slowness of

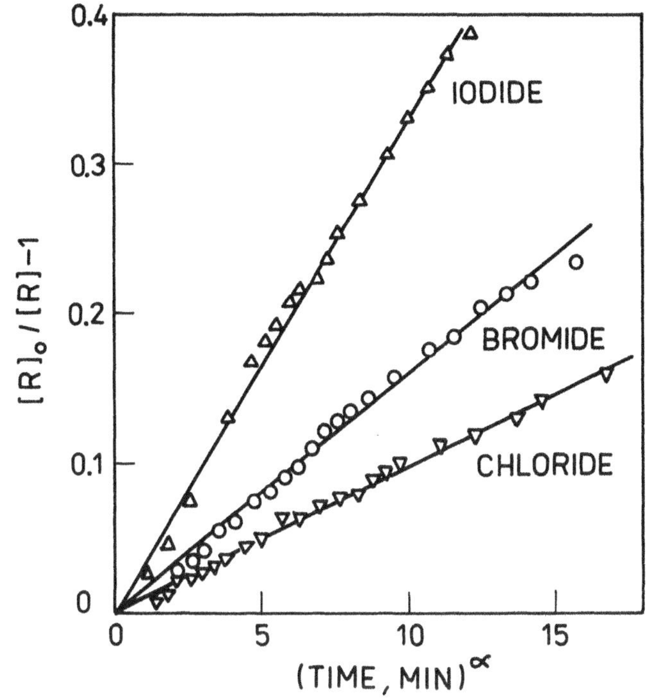

Fig. 2.50. Decay of C_5H_{11} radicals at 77 K in glassy n-$C_5H_{11}Cl$, n-$C_5H_{11}Br$, and n-$C_5H_{11}I$ γ-irradiated at 77 K. Rate parameters in Table 2.6

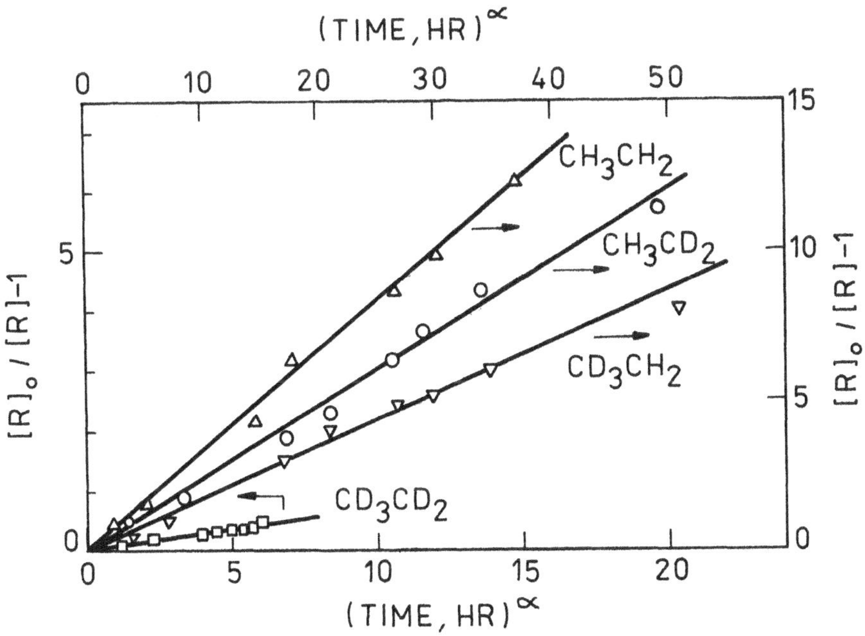

Fig. 2.51. Decay of isotopic ethyl radicals in γ-irradiated at 77 K ethyl iodides. Rate parameters in Table 2.6

n-C_4H_9 relative to the other radicals present as solutes in 3MP-d_{14} illustrates the marked effect of matrix isomerism on decay kinetics. The dominant matrix effect is also seen by comparing n-alkyl radical decay in 3MP-d_{14} with those in parent glassy chlorides.

Basically, the same picture of the kinetic behaviour of alkyl radicals in low-temperature glasses emerges from the extensive studies of Tolkatchev's group. They are presented below in some detail to illustrate the efforts undertaken to arrive at the adequate description of reaction patterns given by the time-dependent rate constant (2.2).

Applicability of $\sqrt{t}$ Law

In 1976 Bol'shakov and Tolkatchev reported the results of ESR studies on the reaction of methyl radicals with methanol

$$CH_3 + CH_3OH \longrightarrow CH_4 + CH_2OH \qquad (2.51)$$

at 77 K. The methyl radicals were produced by photolysis of diphenylamine (3 mol%) followed by dissociative electron capture by methyl

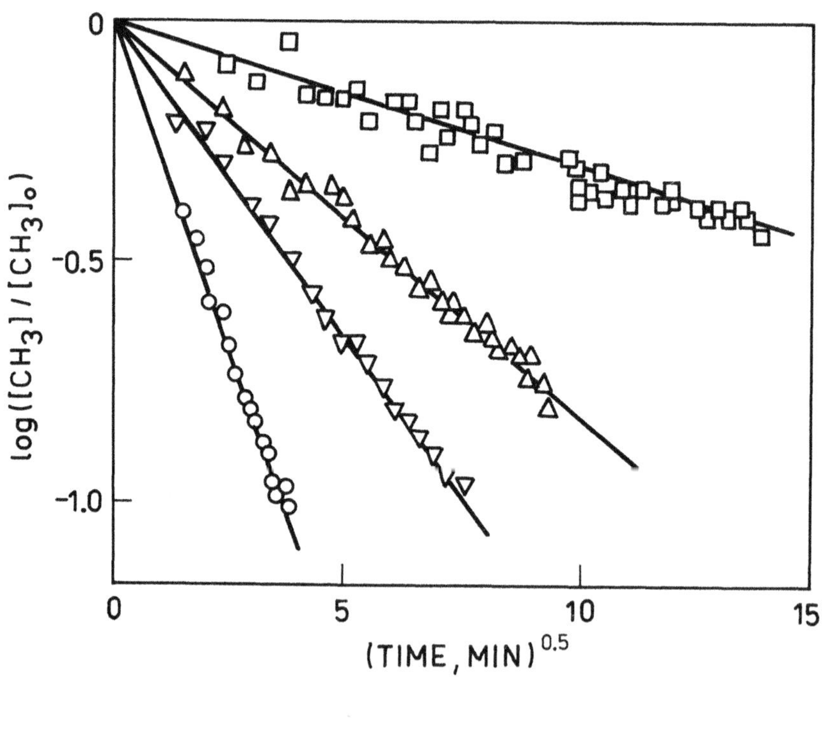

Fig. 2.52. Decay of CH_3 radicals at 77 K in matrices of 18 mol% of D_2O and 82 mol% of methanol with isotopic composition: 100% CH_3OH (○), 67% CH_3OH + 33% CD_3OD (▽), 50% CH_3OH + 50% CD_3OD (△), and 35% CH_3OH + + 65% CD_3OD (□)

iodide (1.5 vol%) in methanol containing varying amounts of H_2O or D_2O. The kinetic curves for CH_3 decay for a fivefold variation of initial concentration did superimpose. They yield a straight line when plotted as $\log([CH_3]/[CH_3]_o)$ vs. $t^{0.5}$ regardless of the isotopic composition of methanol or water content, cf. Figs. 2.52 and 2.53.

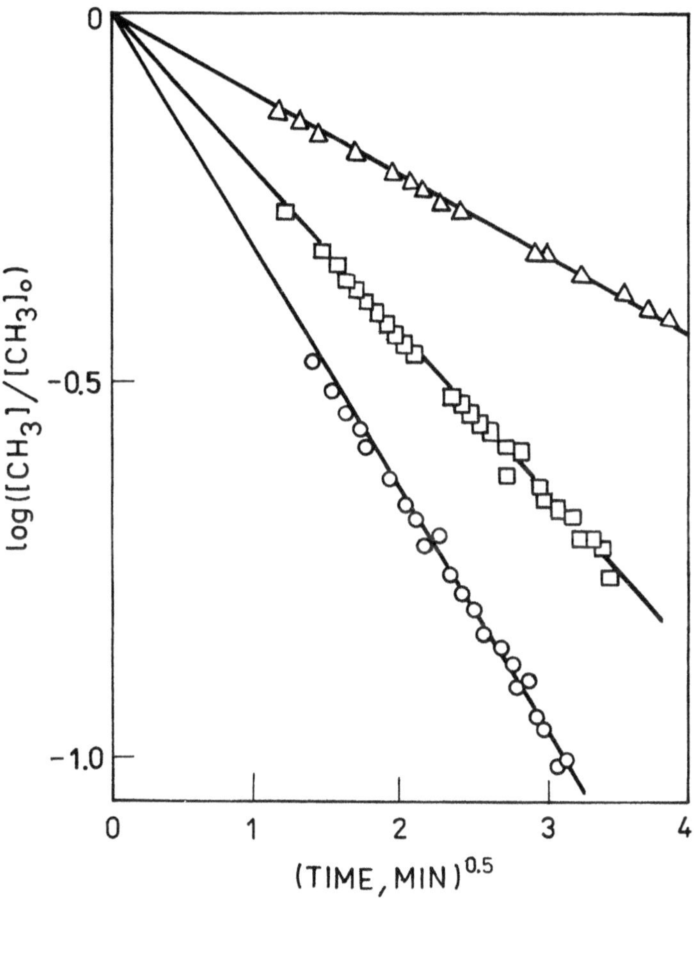

Fig. 2.53. Decay of CH_3 radicals at 77 K in matrices: CH_3OH (○), 69 mol% CH_3OH + 31 mol% H_2O (□), 56 mol% CH_3OH + 44 mol% H_2O (Δ)

The same experimental method was used by Stepanov et al. (1978) to study CH_3 radical reactions with matrix molecules in glasses C_2H_5OH, $(CH_2OH)_2$, n- and iso-C_3H_7OH, n- and iso-C_4H_9OH, C_2D_5OD and iso-C_3D_7OH. In all cases, the methyl radical decay was found to follow eqn. (2.49) with $\alpha = 0.5$, cf. Figs. 2.54 - 2.57 showing the data for two different times of UV-irradiation: 5 and 30 s. As $[CH_3]_o$ the initial concentration after 30 s of irradiation was taken. The same kinetic pattern was

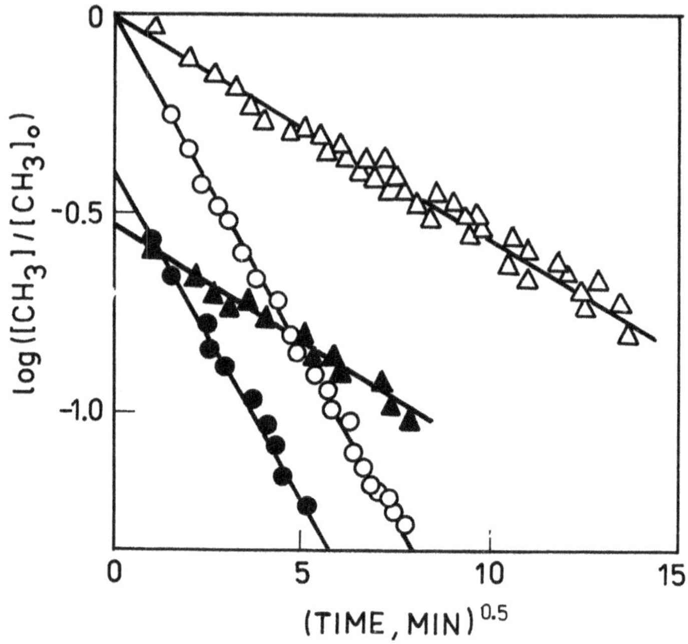

Fig. 2.54. Decay of CH_3 at 77 K in $(CH_2OH)_2$ (△, ▲) and C_2H_5OH (○, ●)

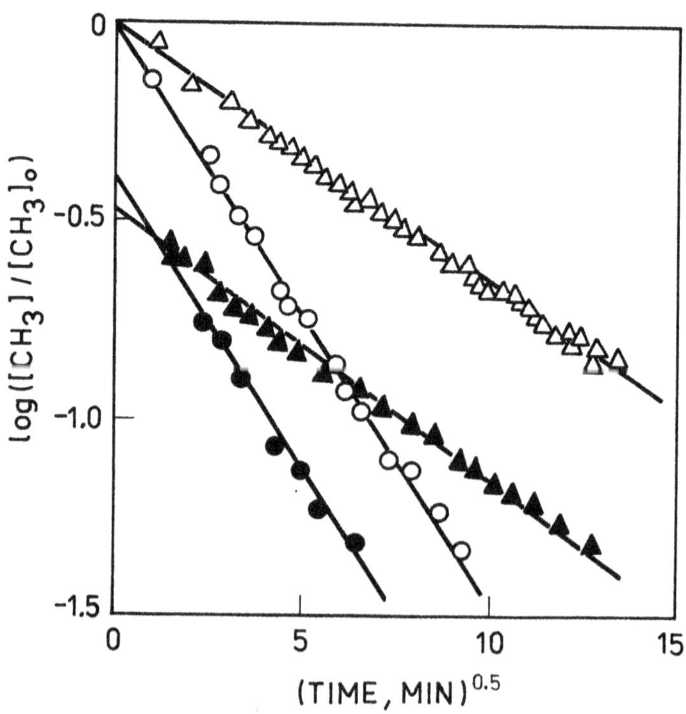

Fig. 2.55. Decay of CH_3 at 77 K in iso- (△, ▲) and n-C_3H_7OH (○, ●)

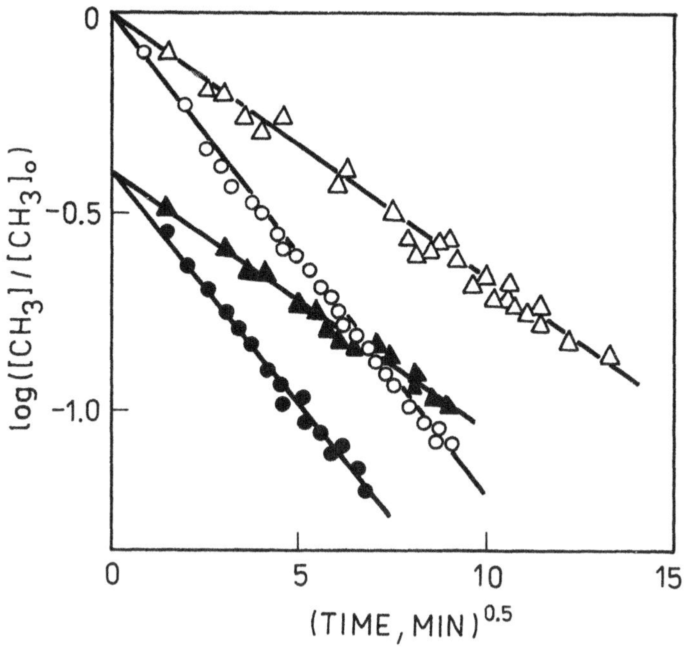

Fig. 2.56. Decay of CH_3 at 77 K in iso- (△, ▲) and n-C_4H_7OH (○, ●)

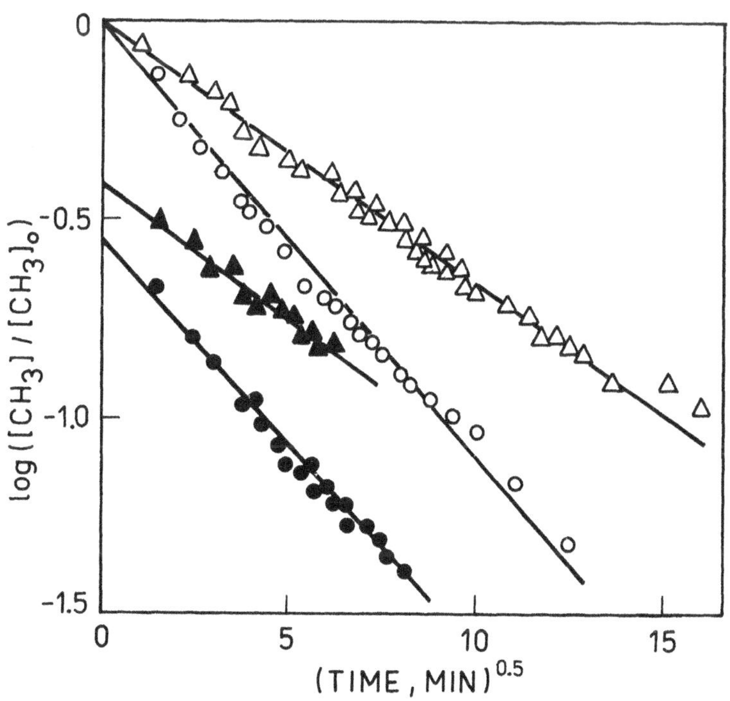

Fig. 2.57. Decay of CH_3 at 77 K in iso- (△, ▲) and n-$C_5H_{11}OH$ (○, ●)

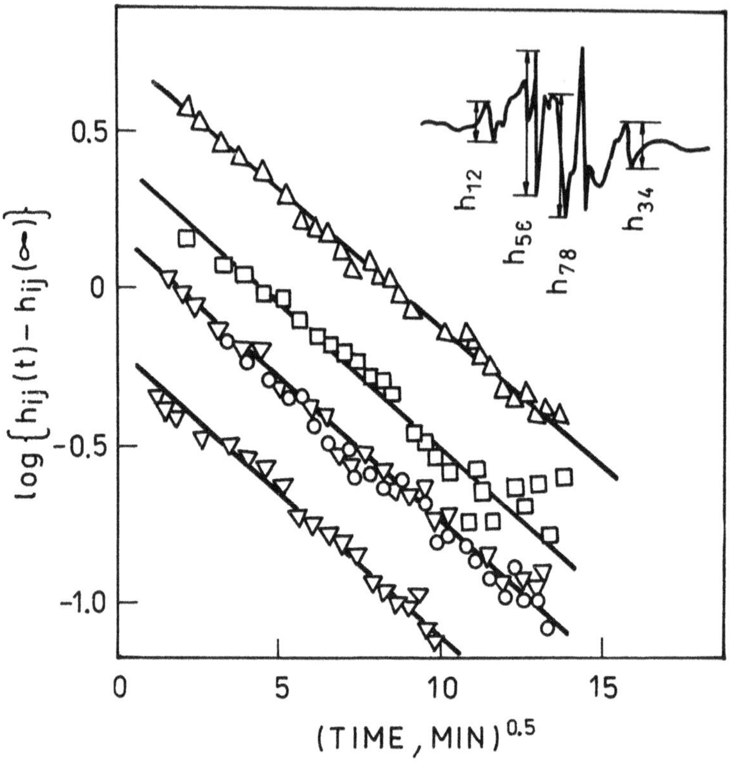

Fig. 2.58. Kinetic curves for C_2H_5 radicals reaction with methanol at 97 K. Insert illustrates the measured components of the ESR spectrum of UV-irradiated methanol-C_2H_5Br-diphenylamine mixture: h_{12} (▽), h_{34} (○), h_{56} (△), h_{78} (□). Different lines with identical points correspond to different initial concentrations

observed for reactions of C_2H_5, C_3H_7, and C_5H_{11} radicals in methanol glasses, cf. Figs. 2.58 - 2.61. The numerical values of rate parameter B are collected in Table 2.7. Thus it seemed that a new kind of kinetic law, the so called $\sqrt{t}$ law, was found for solid-phase reactions and Bol'shakov et al. (1980) had undertaken the experimental verification of the stability of that law for the reaction of H atom abstraction. The points of interest were the thermal treatment of the matrix and the way of radical generation.

Generally the low-temperature investigations of radical reactions are carried out immediately after the matrix has been frozen and one can suppose that the nonexponential kinetics of H atom abstraction reaction is associated with the decrease of the reaction constant with time due to matrix structure relaxation to the equilibrium state. This phenomenon was seen by Neiss and Willard (1975) for CH_3 radical decay in 3-methylpentane glass.

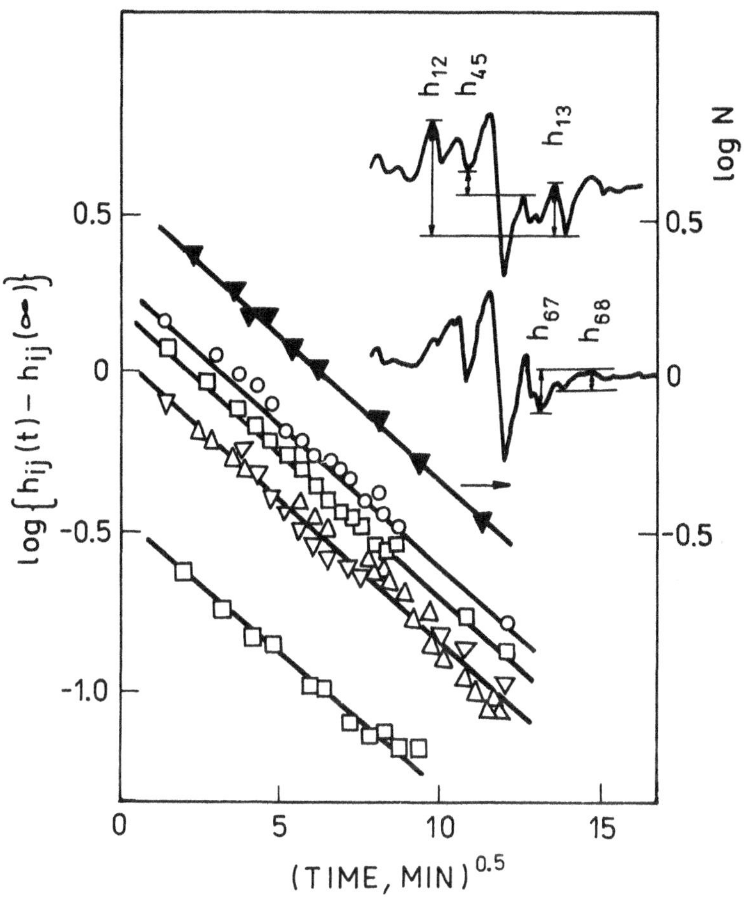

Fig. 2.59. Kinetic curves for C_4H_9 radicals reaction with methanol at 87 K. Insert illustrates the measured components of the ESR spectrum of UV-irradiated methanol-C_4H_9Cl-diphenylamine mixture 4 and 30 min after irradiation: h_{12} (□), h_{13} (▽), h_{45} (○), h_{68} (△), N (▼) denotes the butyl radical concentration in relative units

Figs. 2.62 - 2.64 depict the data obtained by Bol'shakov et al. (1980). There is no effect of sample annealing at the reaction temperature before radical generation on decay kinetics of alkyl radicals in alcohol matrices. Furthermore, the decay kinetics is unaffected by the different ways of free radical generation. However, it was hard to prove the universality of $\sqrt{t}$ law for other reactions of alkyl radicals even in the same kind of matrices as those discussed above.

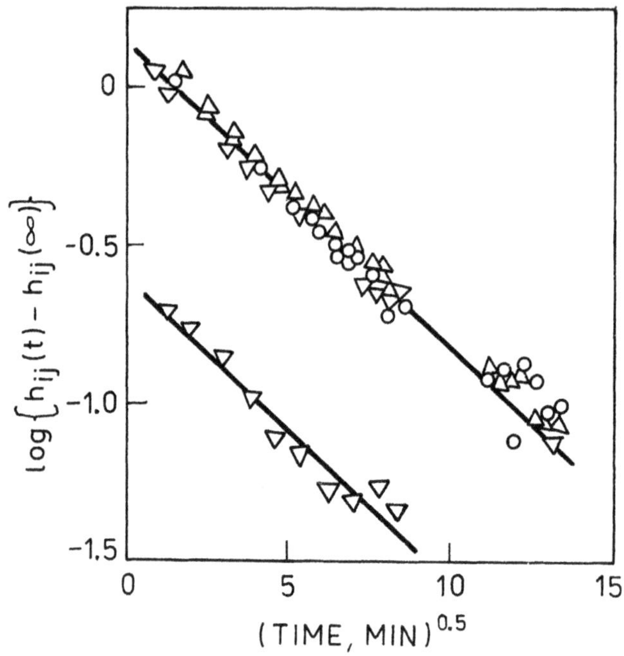

Fig. 2.60. Reaction of C_3H_7 with methanol at 87 K, spectrum components: h_{12} (▲), h_{13} (▼), h_{45} (○), cf. Fig. 2.59

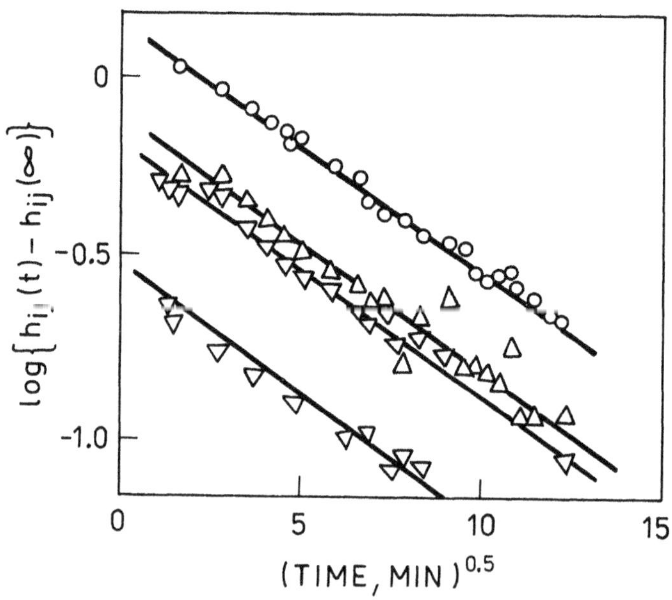

Fig. 2.61. Reaction of C_5H_{11} with methanol at 77 K, spectrum components: h_{12} (▲), h_{13} (▼), h_{45} (○), cf. Fig. 2.59

Table 2.7. Numerical values of rate parameterB for reactions of alkyl radicals with aliphatic alcohol glasses; $\alpha = 0.5$

Reaction	T,K	B, $min^{-1/2}$
$CH_3 + CH_3OH$	77	0.72
$CH_3 + (CH_2OH)_2$	77	0.15
$CH_3 + C_2H_5OH$	77	0.40
$CH_3 + n\text{-}C_3H_7OH$	77	0.34
$CH_3 + n\text{-}C_4H_9OH$	77	0.30
$CH_3 + n\text{-}C_5H_{11}OH$	77	0.26
$CH_3 + iso\text{-}C_3H_7OH$	77	0.15
$CH_3 + iso\text{-}C_4H_9OH$	77	0.16
$CH_3 + iso\text{-}C_5H_{11}OH$	77	0.16
$C_2H_5 + CH_3OH$	87	0.20
$n\text{-}C_3H_7 + CH_3OH$	87	0.21
$n\text{-}C_4H_9 + CH_3OH$	87	0.21
$n\text{-}C_5H_{11} + CH_3OH$	87	0.16

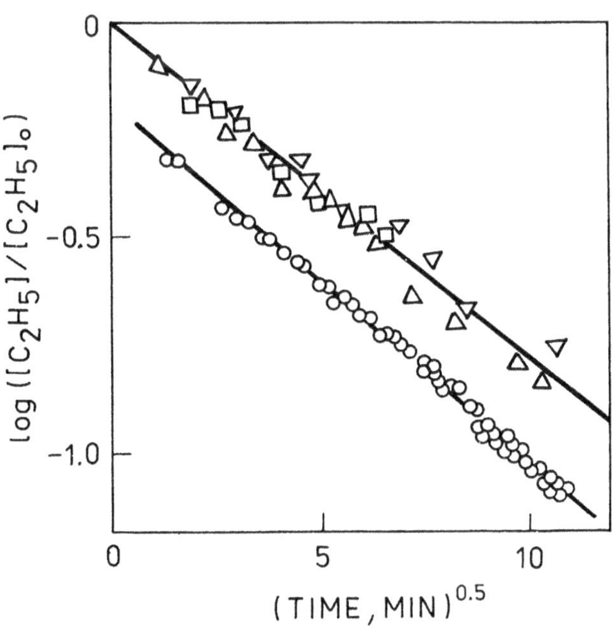

Fig. 2.62. Decay of C_2H_5 at 87 K in methanol annealed for 0 (Δ, O), 14 (∇), and 21 (□) days; generation from C_2H_5I (Δ, ∇, □), from C_2H_4 (O)

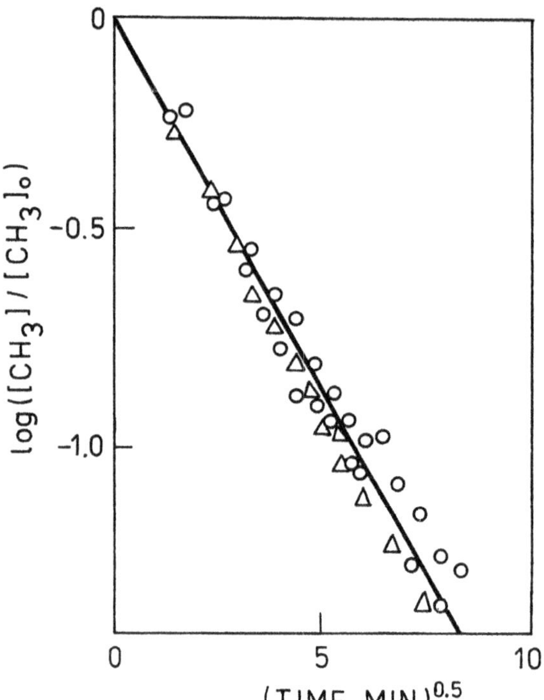

Fig. 2.63. Decay of methyl radicals in ethanol at 77 K: sample annealed for 6 days (△), sample cooled gradually to 77 K over a two-day span (○)

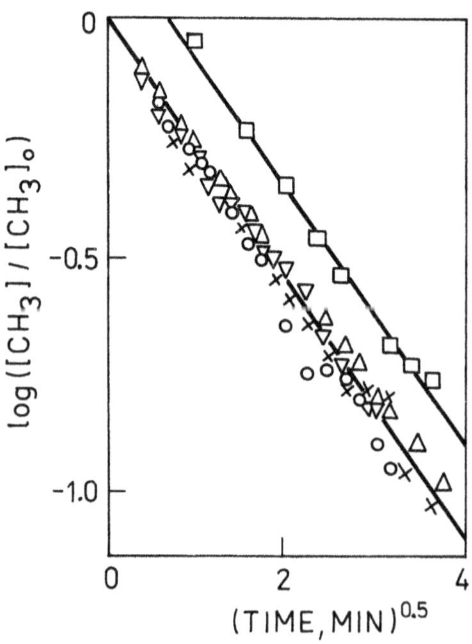

Fig. 2.64. Decay of methyl radicals in methanol at 77 K: sample annealing for 6 (○), 12 (△), 26 (▽), and 44 days (×); immediately after sample freezing (□); generation technique: from CH_3I (○, △, ▽, ×), from CH_2OH (□)

Oxidation Kinetics

For the reaction of radical oxidation there was evidenced the effect of annealing on reaction kinetics, cf. the results of Zaskul'nikov and Tolkatchev (1982) depicted in Fig. 2.65.

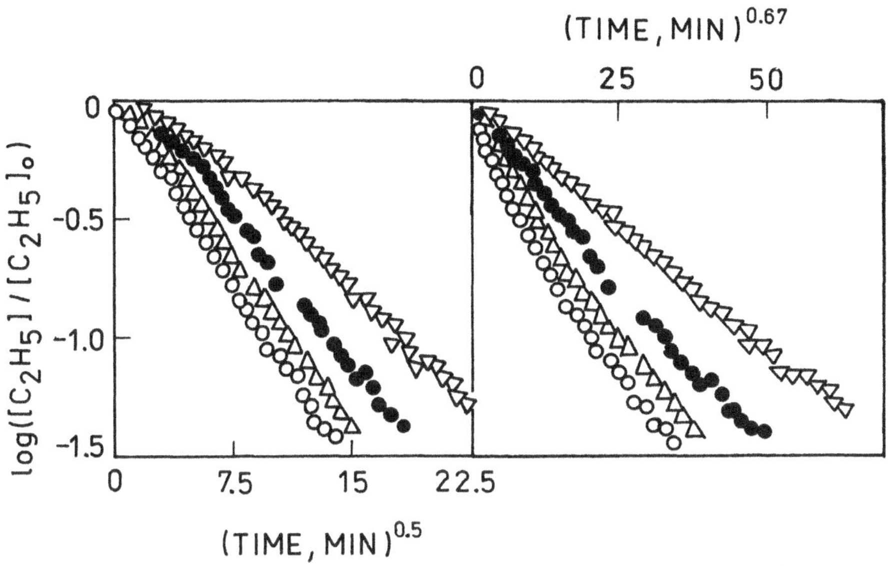

Fig. 2.65. Oxidation of ethyl radicals in methanol at 77 K: unannealed sample (○), sample annealed for 10 min (△), 3.3 hrs (●), and 40 hrs (▽)

Table 2.8. Numerical values of $(B/\alpha)/[O_2]$, cf eqn. (2.52), for radical oxidation in non-annealed methanol-d_4 matrix at 87 K; $\alpha = 0.5$

Radical	$(B/\alpha)/[O_2]$ $10^{-2} min^{-1/2} atm^{-1}$
CH_3	3.4 ± 0.25
C_2H_5	3.7 ± 0.1
n-C_4H_9	3.25 ± 0.35
iso-C_4H_9	3.9 ± 0.15
C_6H_{11}	3.4 ± 0.45
sec-C_4H_9	3.45 ± 0.1
n-C_6H_{13}	3.3 ± 0.7
tert-C_4H_9	4.15 ± 0.25
$C_6H_5CH_2CH_2$	2.9 ± 0.45
$C_6H_5CH_2$	2.8 ± 0.35
$CH_3C_6H_4CH_2$	2.4 ± 0.45

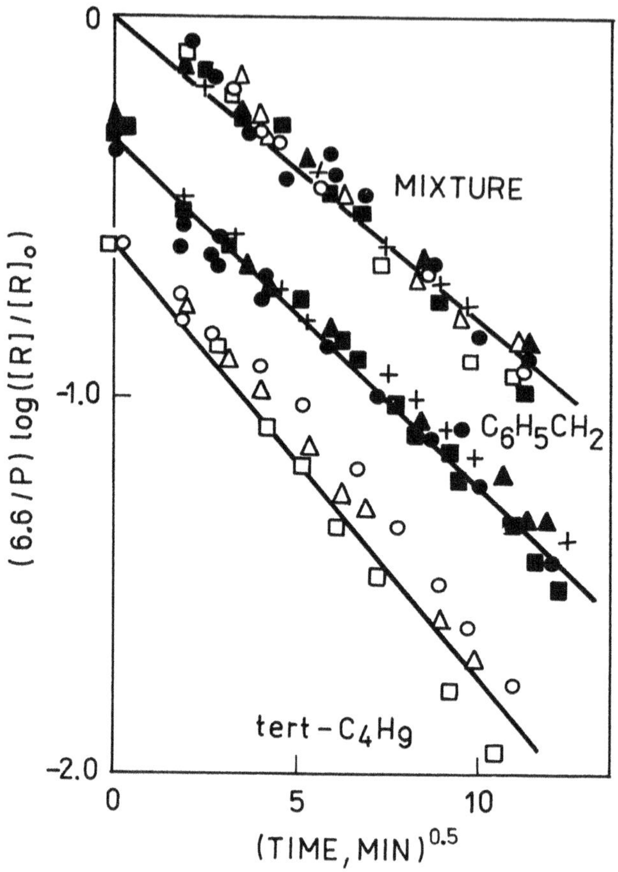

Fig. 2.66. Oxidation of tert-C_4H_9 (empty dots) and $C_6H_5CH_2$ (full dots) radicals in non-annealed methanol-d_4 matrix at 87 K. Oxygenation pressure: 1.1 (○), 3.0 (Δ), 6.1 (+), 6.6 atm. (□). The curves for tert-C_4H_9 and $C_6H_5CH_2$ radicals are displaced downwards from the curve labelled "mixture" which illustrates oxidation kinetics of both kinds of radicals in the same sample

It was reported that annealing changes the value of α from 1/2 to 2/3, the latter value being the new law. Further extensive studies in the same Laboratory resulted in full acceptance of the form (2.2) of the time-dependent rate constant with $0 < \alpha \leq 1$. It is seen in Bagryansky et al. (1983) studies on oxidation kinetics of CH_3, C_2H_5, n-C_4H_9, iso-C_4H_9, sec-C_4H_9, tert-C_4H_9, n-C_6H_{11}, C_6H_5-CH_2, CH_3-C_6H_4-CH_2, and $C_6H_5CH_2CH_2$ radicals in methanol matrix at 87 K. The kinetics of oxidation was described with the pseudo first-order equation, which in our notation reads

$$\ln([R]/[R]_0) = -(B/\alpha)t^{\alpha}[O_2] \qquad (2.52)$$

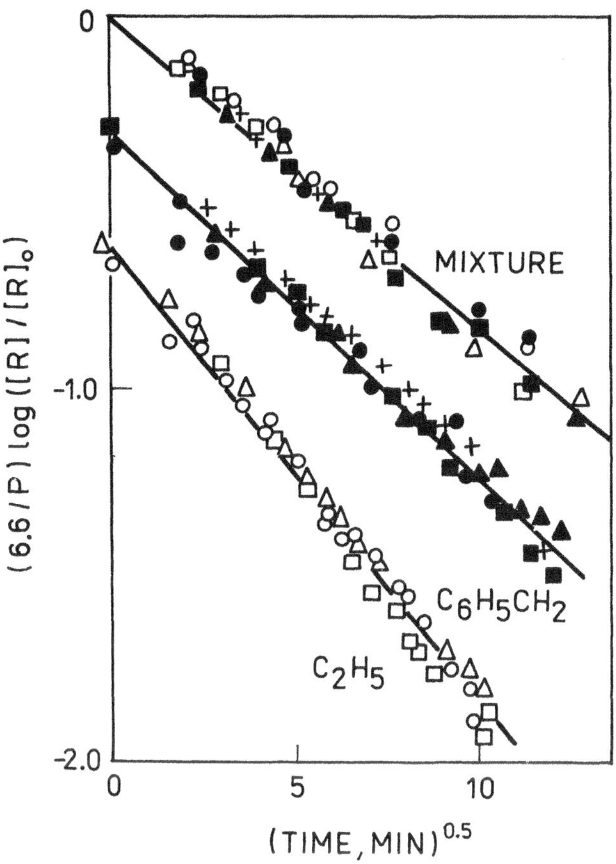

Fig. 2.67. Oxidation of C_2H_5 (empty dots) and $C_6H_5CH_2$ (full dots) radicals in non-annealed methanol-d_4 matrix at 87 K. Oxygenation pressure: 1.1 (○), 3.0 (△), 6.1 (+), and 6.6 atm. (□). The curves for C_2H_5 and $C_6H_5CH_2$ radicals are displaced downwards from the curve labelled "mixture" which illustrates oxidation kinetics of both kinds of radicals in the same sample

According to eqn. (2.52) the kinetic curves have to coincide for various oxygenation pressures, P, in the coordinate systems $(1/P)\log([R]/[R]_o)$ vs. t^{α}. The parameter α was reported to differ in experiments with different radicals, being equal to 0.53±0.03 for ethyl radicals, 0.45±0.03 for benzyl radicals, and 0.58±0.03 for tert-butyl radicals. These cases are illustrated in Figs. 2.66 and 2.67 in which unified coordinates $(6.6/P)\log([R]/[R]_o)$ vs. $t^{1/2}$ were used due to small differences of α from 0.5.

Fig. 2.68 illustrates the oxidation kinetics of benzyl and tert-butyl radicals in methanol-d_4 matrix annealed for 1000 min at the reaction temperature before radical generation. It is seen to be fitted quite

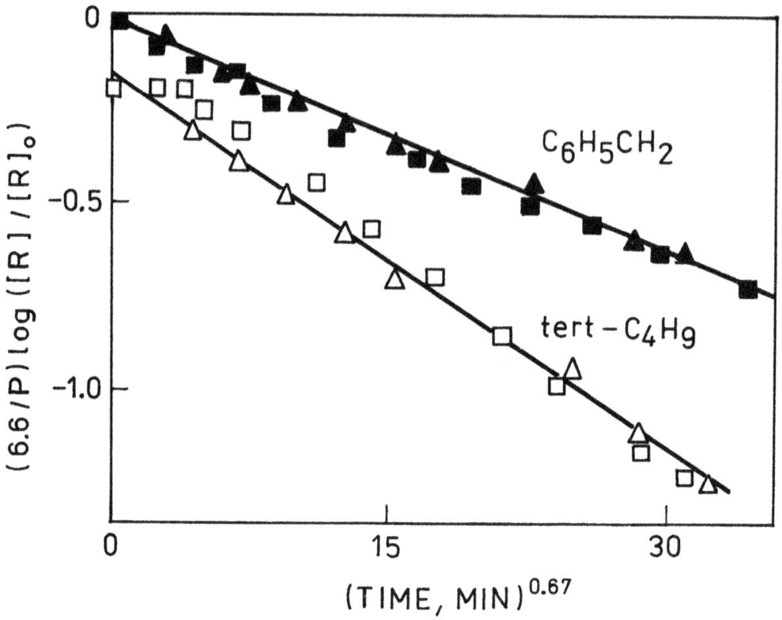

Fig. 2.68. Oxidation of $C_6H_5CH_2$ and tert-C_4H_9 radicals in methanol-d_4 matrix annealed before irradiation for 1000 min at 87 K. Oxygenation pressure: 3.0 (△), 6.6 (□) atm. The curve for tert-C_4H_9 radicals is displaced downwards.

well by α equal to 2/3.

Looking at the numerical values of $(B/\alpha)[O_2]$ in eqn. (2.52), cf. Table 2.8, for radical oxidation in non-annealed methanol-d_4 matrix at 87 K one finds that they differ somewhat. To clarify the origin of the differences Bagryansky et al. (1983) studied the oxidation kinetics for two different radicals obtained in the same sample by adding two halides. They used pairs of radicals for which the ESR lines do not overlap and the values of (B/α) differ significantly, cf. Fig. 2.66 for tert C_4H_9 and $C_6H_5CH_2$ radicals, and Fig. 2.67 for C_2H_5 and $C_6H_5CH_2$ radicals. It is seen that if the kinetics were studied in each sample separately the oxidation rates were found to be different. If the kinetics were studied in mixed samples the oxidation rates were found to coincide practically. Thus it was concluded that small differences in the numerical values of $(B/\alpha)/[O_2]$ for all the studied radicals were probably associated with the effects of the halides upon matrix structure. The fact that the oxidation rate does not depend on chemical properties of the radicals was taken as the evidence that the reaction step for these systems is not a limiting step in the meaning attributed to this term in liquid-phase reactions. It was suggested that an elementary act in solids is of complex cooperative character and involves several coordination spheres around the reagents. From this point of view it would be clear why the oxidation rate is independent of such parameters as steric factors, reaction radii, etc. As to the kinetic aspects of the problem Bagryansky et al. (1983) have shown that retarded oxidation kinetics can be described in terms of a time-dependent rate constant (2.2). Matrix relaxation has been shown to affect this kinetics, however, to express its contribution to the rate constant as a multiplier in eqn. (2.52), not discussed here, one would have to accept at first some universality of the $t^{2/3}$ law.

SOME OTHER SPECIES

To end the discussion on reaction kinetics in glassy media it is worth mentioning some other species, such as Cl_2^- anion-radicals in the above discussed system $HCl-H_2O-ZnCl_2$, NO_3 in $KNO_3-Ca(NO_3)_2$ glass and especially arylcarbenes in various low-temperature glasses.

Disproportionation of Cl_2^-

In the above discussed system $HCl-H_2O-ZnCl_2$ beside the decay of trapped hydrogen atoms the bimolecular reaction

$$Cl_2^- + Cl_2^- \longrightarrow Cl_2 + 2Cl^- \qquad (2.53)$$

was also followed (Szajdzinska-Pietek et al. 1982).

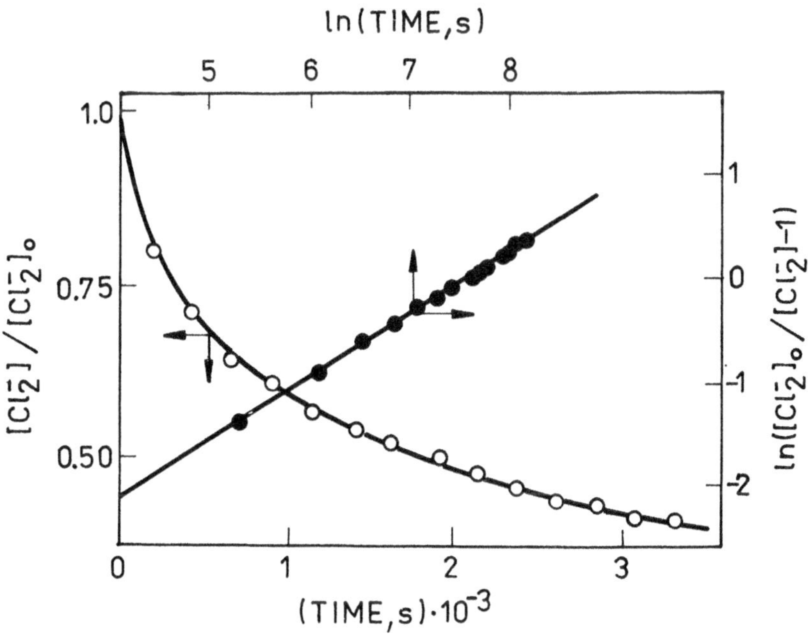

Fig. 2.69. Kinetics of Cl_2^- decay at 140 K in $HZn_2Cl_5\cdot 4H_2O$ γ-irradiated at 77 K

The change of reaction order was observed for $HZn_2Cl_5\cdot 4H_2O$, matrix I, near the RTL peak C at 151 K, cf. Fig. 2.69 showing the second-order decay with time-dependent rate constant (2.2) at 140 K and Fig. 2.70 showing the respective first-order decay at 150 K. Thus at temperatures far from that of the proper RTL peak the decay due to bimolecular reaction obeys the second-order kinetic equation while in the vicinity of RTL maximum the process is of the first-order.

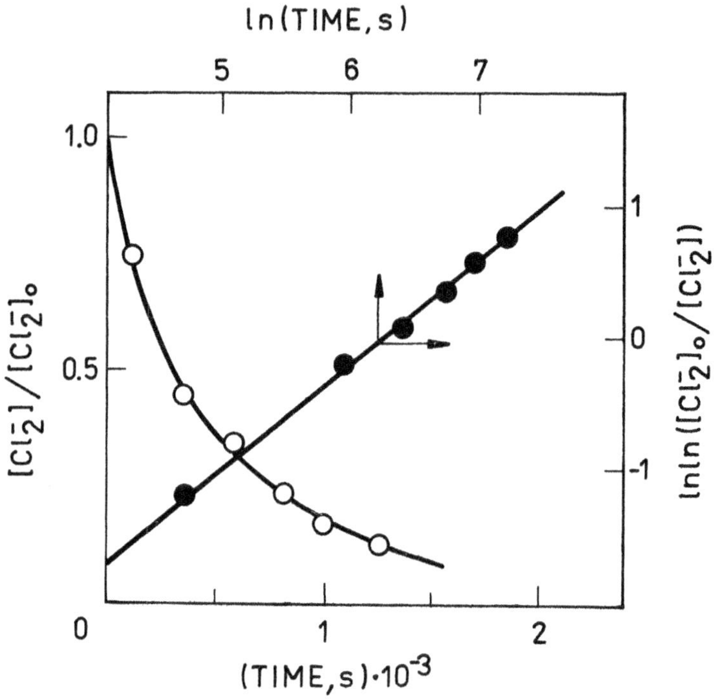

Fig. 2.70. Kinetics of Cl_2^- decay at 150 K in $HZn_2Cl_5 \cdot 4H_2O$ γ-irradiated at 77 K

Recombination $NO_3 + NO_2^{2-} \longrightarrow 2\ NO_3^-$

Hamill (1981) analyzed the data of Barkatt et al. (1972) for recombination

$$NO_3 + NO_2^{2-} \longrightarrow 2\ NO_3^- \tag{2.54}$$

in pulse irradiated 3:2 KNO_3-$Ca(NO_3)_2$ glass from 77 to 486 K in the time range 10^{-6} - 10^2 s. In the temperature range 77 - 393 K the data were found to fit the second-order equal concentration kinetic equation with time-dependent rate constant (2.2), cf. Fig. 2.71. This fully supports the conclusion reached above that if species are generated at the temperature at which the decay is followed the second-order decay pattern is observed for bimolecular reaction.

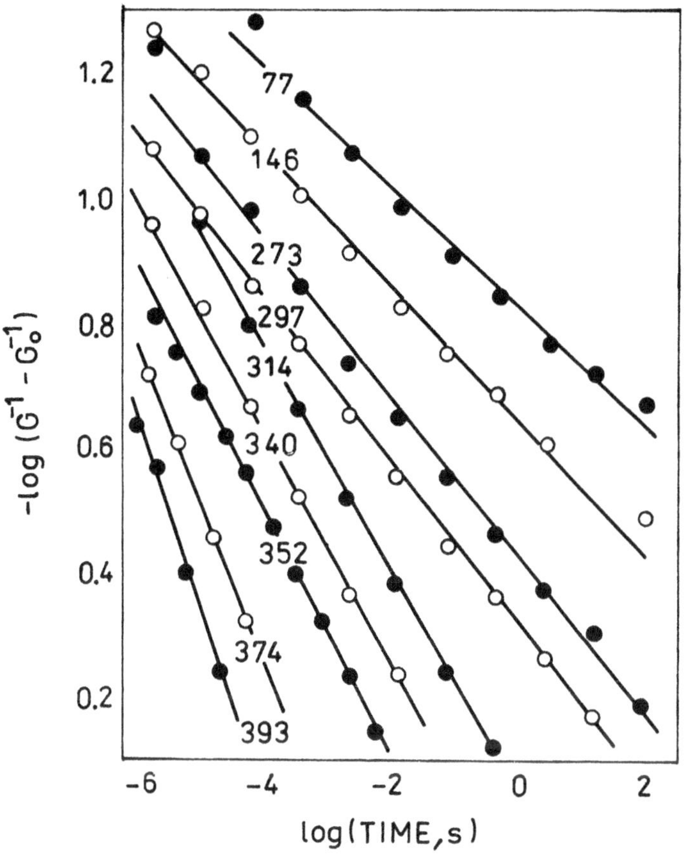

Fig. 2.71. Kinetics of NO_3 decay at 77 - 393 K in pulse-irradiated 3:2 KNO_3-$Ca(NO_3)_2$ glass. Glass transition temperature 341 K

It was stressed (ibid.) that time-dependence of k(t) as given by eqn. (2.2) provides the description of the dependence of the activation energy on time, E(t). Indeed one may have

$$k(t) = A_o \exp\left\{-\left[E_o + \gamma \ln(t/\tau)\right]/kT\right\} \tag{2.55}$$

with

$$\gamma/kT = 1 - \alpha \tag{2.56}$$

since

$$\exp\left[(\alpha - 1)\ln(t/\tau)\right] = (t/\tau)^{\alpha - 1} \tag{2.57}$$

and

$$k(t) = A_o (t/\tau)^{\alpha-1} \exp(-E_o/kT) \, . \qquad (2.58)$$

The lines in Fig. 2.71 are reported to converge at about 10^{-3} s. Consequently τ is equal to about 10^{-13}, E_o to zero, and A_o is temperature independent. One may say e.g. that the activation energy E(t) increases from zero at a paramagnetic 10^{-13} s to 0.7 eV at 10^2 s.

Arylcarbenes Decay

The kinetics of arylcarbene bimolecular reactions in glasses and crystals was followed by the ESR method, cf. Senthilnathan and Platz (1980, 1981). The kinetics were found to be sensitive to the chemical nature of the matrix, the viscosity of the matrix, the concentration of diazo compound, and the history of the sample with respect to photolysis. One effect of the matrix was to cause the nonexponential decay of the ESR signal, I. The experimental data were well fitted by either log I vs. $t^{1/2}$, Fig. 2.72 or log I vs. $t^{1/3}$ dependence, Fig. 2.73. The Arrhenius parameters and isotope effects indicate that hydrogen atom tunnelling reactions are the major decay pathways in the low-temperature matrices.

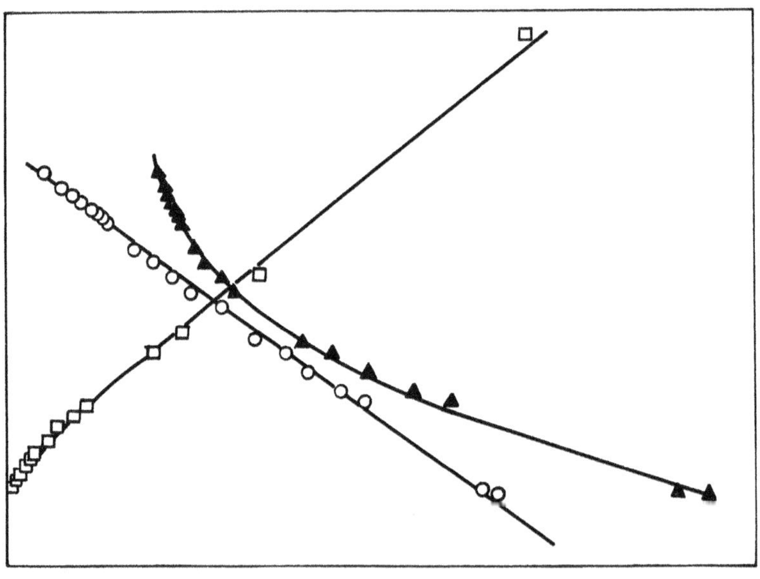

Fig. 2.72. Plots of log I vs. $t^{1/2}$ (○), log I vs. t (▲), 1/I vs. t (□) for diphenylcarbene in 2:3 (iso)propanol (1- and 2-) at 98 K with I_o = 65 and I_∞ = 18 in arbitrary units at the end of 60 min

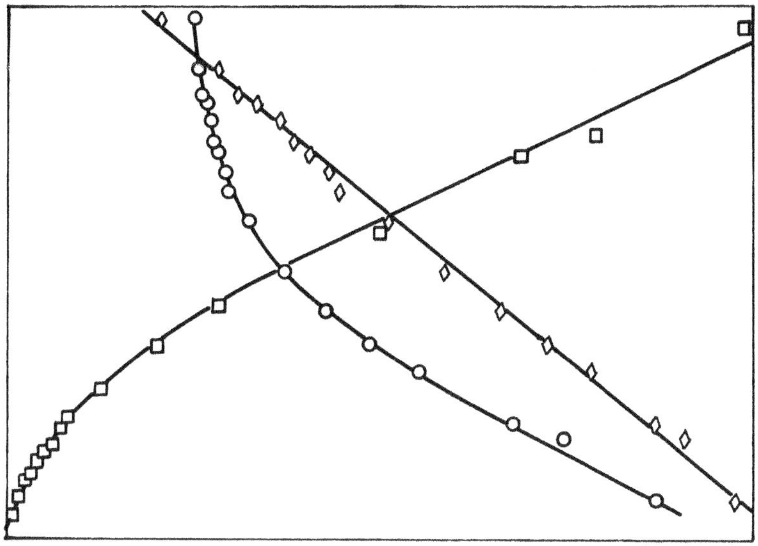

Fig. 2.73. Plots of log I vs. $t^{1/3}$ (◊), log I vs. t (○), 1/I vs. t (□) for diphenyl carbene triplet ESR $\Delta m = 2$ transition in 1:1 C_6F_6/toluene at 98 K with $I_o = 42$ and $I_\infty = 10$ in arbitrary units at the end of 100 min

REFERENCES

Angell CA, Sare EJ (1970) Glass-forming composition regions and glass transition temperatures for aqueous electrolyte solutions. J Chem Phys 52: 1058-1068

Bagdasar'yan KhS, Milutinskaya RI, Kovalev YuV (1967) Luminescence at low temperatures of γ-irradiated organic glasses (Russ). Khim Vys Energ 1: 127-135

Bagryansky VA, Zaskul'nikov VM, Tolkatchev VA (1983) The kinetics of the retarded reaction of free radicals with oxygen in methanol glass. Chem Phys 78: 41-48

Barkatt A, Angell CA (1978) Use of structural probe ions for relaxation studies in glasses. 2. Temperature-jump and temperature-ramp studies of cobalt(II) in nitrate glasses. J Phys Chem 82: 1972-1979

Bartczak WM, Hilczer M, Kroh J (1981) Trapped electron in frozen ionic solutions - a theoretical model. Radiat Phys Chem 17: 431-437

Bartczak WM, Kroh J, Stradowski C (1977) Kinetic model for electron tunnelling in frozen irradiated systems. J Chem Phys 66: 2737-2741

Baxendale JH, Sharpe PHG (1976) A pulse radiolysis study of electrons in 1-propanol at low temperatures. Int J Radiat Phys Chem 8: 621-624

Benderskii VA, Krivenko AG, Rukin AN (1980) Anomalous mobility of hydrogen and deuterium atoms in aqueous solutions of electrolytes (Russ). Khim Vys Energ 14: 400-405

Bhattacharya D, Willard JE (1981) Radiolytic production of trapped hydrogen atoms from organic compounds in Xe, Kr, and Ar at 10 K. J Phys Chem 85: 154-159

Bol'shakov BV, Stepanov AA, Tolkatchev VA (1980) The $\sqrt{t}$ law in solid phase reactions. Verification of the $\sqrt{t}$ law stability in H-atom abstraction reactions. Int J Chem Kin 12: 271-281

Bol'shakov BV, Tolkatchev VA (1976) A kinetic peculiarity of free valence migration in the solid state as studied through the reaction of methyl radicals with methanol at 77 K. Chem Phys Lett 40: 468-470

Buxton GV, Gillis HA, Klassen NV (1976) Evidence for a second kind of trapped electron in some deuterated aqueous glasses at low temperatures. A pulse radiolysis study. Can J Chem 54: 367-381

Claridge RFC, Willard JE (1965) Production of trapped radicals from alkyl halides in organic glasses by dissociative electron attachment and by photodissociation. J Am Chem Soc 87: 4992-4997

Dainton FS, Pilling MJ, Rice SA (1975) Theoretical model of electron scavenging in irradiated glassy media based on a tunnelling mechanism. J Chem Soc, Faraday Trans II 71: 1311-1321

DeBolt MA, Easteal AJ, Macedo PB, Moynihan CT (1976) Analysis of structural relaxation in glass using rate heating data. J Am Ceram Soc 59: 16-21

Debye P, Edwards JD (1952) Long-lifetime phosphorescence and the diffusion process. J Chem Phys 20: 236-239

French WB, Willard JE (1968) Radical decay kinetics in organic glasses. Spatial effects and isotope effects. J Phys Chem 72: 4604-4608

Funabashi K, Hamill WH (1978) A random-walk model for spectral relaxation of trapped electrons in alcohols. Chem Phys Lett 56: 175-177

Funabashi K, Hamill WH (1979) The role of electron - phonon interaction and non-Gaussian transport in spectral changes of trapped electrons in glasses. Can J Chem 57: 197-206

Gillis HA, Teather GG, Buxton GV (1978) Pulse radiolysis of deuterated aqueous lithium chloride glasses. Dependence of the yields and rates of reaction of visible and infrared absorbing trapped electrons on lithium chloride concentration. Can J Chem 56: 1889-1897

Gillis HA, Walker DC (1976) Luminescence and infrared absorption from laser excitation of trapped electrons in aqueous glasses. J Chem Phys 65: 4590-4595

Hagekyriakou J, Fleming RJ (1982) Determination of the kinetic order of thermoluminescence in the presence of a distribution of electron trap activation energies. J Phys D, Appl Phys 15: 163-170

Hamill WH (1978) Charge-pair distribution and recombination kinetics. The role of non-Gaussian diffusion. J Phys Chem 82: 2073-2077

Hamill WH (1981) Temperature dependence of dispersive hopping in recombination of charge pairs in disordered solids. Chem Phys Lett 77: 467-469

Hamill WH, Funabashi K (1977) Kinetics of electron trapping reactions in amorphous solids; a non-Gaussian diffusion model. Phys Rev B 16: 5523-5527

Hodge JM, Angell CA (1978) Ionic hydration and secondary relaxations in vitrified concentrated aqueous solutions. J Phys Chem 82: 1761-1766.

Johari GP, Jones SJ (1975) Study of the low-temperature "transition" in ice I_h by thermally stimulated depolarization measurements. J Chem Phys 62: 4213-4223

Klassen NV, Adams RJ, Teather GG, Ross CK (1980) the reaction of e^-_{vis}

and e^-_{tr} with Cl_2^- in $LiCl/D_2O$ glasses as measured from 10^{-7} to 2000 s. J Phys Chem 84: 3609-3613
Klassen NV, Gillis HA, Teather GG, Kevan L (1975) Pulse radiolysis studies of time dependent spectral shifts of the solvated electron in ethanol and deuterated ethanol glasses at 76 K.J Chem Phys 62: 2474-2478
Konigshang M, Moan J, Bøcman B, Høvik HH (1977) Photobleaching of radiation induced trapped electrons in neutral and alkaline glasses. Hydrogen atom formation. Chem Phys Lett 47: 255-258
Kroh J, Plonka A (1975) Trapped hydrogen atom decay in γ-irradiated sulfuric acid glasses at 63 - 90 K. J Phys Chem 79: 2600-2602
Kroh J, Stradowski C (1975) Electron tunnelling in frozen matrices. Int J Radiat Phys Chem 7: 23-28
Leeuwen van JW, Heijmann MGJ, Nauta H, Casteleijn G (1980) A pulse radiolysis study of electron tunneling in a 8 M NaOH glass between 4 and 100 K. J Chem Phys 73: 1483-1489
Lefik W, Plonka A, Kroh J (1986) On kinetic interpretation of post-irradiation decay of trapped electrons. J Nucl Radioanal Chem - in press
May R, Walker DC (1978) Inhomogeneous bleaching of trapped electron absorption bands in aqueous glasses during laser excitation. J Chem Soc, Faraday Trans II 74: 1833-1943
Miller JR (1975) Reactions of trapped electrons by quantum mechanical tunneling observed by pulse radiolysis of an aqueous glass. J Phys Chem 79: 1070-1078
Miller JR, Clifft BE, Hines JN, Runowski RF, Johnson KW (1976) Solvations of electrons in alcohol glasses from 10^{-6} to 10^2 s after pulse radiolysis at 77 K. J Phys Chem 80: 457-459
Miyazaki T, Azuma N, Fueki K (1984) Kinetic behavior of H and D atoms γ-irradiated SiO_2 as studied by ESR spectroscopy. J Am Ceram Soc 67: 99-102
Montroll EW, Bendler JT (1984) On Lévy (or stable) distributions and Williams-Watts model of dielectric relaxation. J Statist Phys 34: 129-162
Neiss MA, Willard JE (1975) Effects of solutes, deuteration, and annealing on the production and decay of radicals in γ-irradiated 3-methylpentane glasses. J Phys Chem 79: 783-794
Neta P, Fessenden RW, Schuler RH (1971) An electron spin resonance study of rate constants for reaction of hydrogen atoms with organic compounds in aqueous solutions. J Phys Chem 75: 1654-1666
Nguen TQ, Walker DC (1977) Luminescence following laser excitation of trapped electrons in mixed-solute aqueous glasses. J Chem Soc, Faraday Trans I 73: 1958-1971
Ogasawara M, Kevan L (1978) Laser photoionization study of time dependent spectral shifts of localized electrons in ethanol glass. Temperature effects. J Phys Chem 82: 378-382
Onsager L, Staebler DL, Mascarenhas S (1978) Electrical effects during condensation and phase transitions of ice. J Chem Phys 68: 3823-3828
Plonka A (1981) Changes of ESR parameters of trapped hydrogen atoms in sulfuric acid glass stored at 77 K. Radiat Phys Chem 17: 173-175
Plonka A (1983) Mechanism of hydrogen atom decay in γ-irradiated sulfuric acid glasses. Radiat Phys Chem 21: 405-408
Plonka A, Bogus W (1980) Changes of trapped hydrogen atom environment in sulfuric acid glasses at 77 K. Radiat Phys Chem 16: 365-367

Plonka A, Kevan L (1976) Differential saturation study of radial and angular modulation mechanisms of electron spin-lattice relaxation for trapped hydrogen atoms in sulfuric acid glasses. Chem Phys Lett 43: 489-492
Plonka A, Kowalski TS (1977) Thermal decay of paramagnetic species generated in 15 M LiCl aqueous glass by γ-irradiation at 77 K. Bull Acad Polon Sci, Sér Sci Chim 25: 885-890
Plonka A, Kowalski TS, Lefik W, Kroh J (1981) Kinetics of photostimulated conversion of trapped electrons into trapped hydrogen atoms in 15 M LiCl glasses. Radiat Phys Chem 17: 375-377
Plonka A, Kroh J, Wyszywacz K (1978) Kinetic behaviour of trapped hydrogen atoms in sulfuric acid glasses γ-irradiated at 63 - 90 K. Radiochem Radioanal Lett 36: 257-263
Plonka A, Lefik W, Kroh J (1979) Photostimulated decay of trapped electrons in alkaline ice. Pseudo first-order kinetics with time-dependent rate constant. Chem Phys Lett 62: 271-273
Plonka A, Szajdzinska-Pietek E (1980) Thermal decay of paramagnetic species generated by γ-irradiation at 77 K of zinc chloride/hydrochloric acid glasses. Bull Acad Polon Sci, Sér Sci Chim 28: 53-60
Plonka A, Wypych M, Lefik W, Kroh J (1984) Radioluminescence kinetics of frozen solutions of LiCl. Radiat Phys Chem 23: 715-718
Rice SA (1975) Conjectures on the structure of amorphous solid and liquid water. In: Topics in Current Chemistry 60: 109-200. Springer, Berlin Heidelberg New York
Rice SA, Kevan L (1977) Comparison of photoconductivity and optical spectra of the trapped electron in polar aqueous and alcoholic glasses. J Phys Chem 81: 847-850
Roy CR, Willard JE (1972) Further evidence of free radical decay in γ-irradiated alkyl halide and hydrocarbon-alkyl halide glasses. J Phys Chem 76: 1405-1410
Senthilnathan VP, Platz MS (1980) Determination of the absolute rates of decay of arylcarbenes in various low temperature matrices by electron spin resonance spectroscopy. J Am Chem Soc 102: 7637-7643
Senthilnathan VP, Platz MS (1981) Conformational barriers in triplet 1 and 2-naphtylcarbene. 2. Absolute rate of decay of arylcarbenes by electron spin resonance spectroscopy. J Am Chem Soc. 103: 5503-5511
Shiraishi H, Kadol H, Katsumara Y, Tabata Y, Oshima K (1976) Electron spin resonance studies on hydrogen atoms formed in pure and acidic ices under electron irradiation. Motional narrowing and electron spin polarization effect. J Phys Chem 80: 2400-2406
Sprague ED (1973) Hydrogen atom abstraction by methyl radicals in 3-methylpentane glass at 77° K. J Phys Chem 77: 2066-2070
Sprague ED, Schulte-Frohlinde D (1973) Electron spin resonance investigation of the disappearance of trapped hydrogen atoms in γ-irradiated sulfuric acid glasses. J Phys Chem 77: 1222-1225
Stepanov AA, Tkatchenko VA, Bol'shakov BV, Tolkatchev (1978) The kinetics of alkyl radical reactions with aliphatic alcohol glasses as studied by electron spin resonance. Int J Chem Kin 10: 637-648
Szajdzinska-Pietek E, Kroh J, Plonka A (1982) Kinetic patterns of bimolecular reactions. H atom and Cl_2^- radical decay in γ-irradiated zinc chloride/hydrochloric acid glasses. Radiat Phys Chem 20: 135-139
Vacek K (1971) Hydrogen atoms as studied by ESR. Prog Probl Contemp Radiat Chem Proc Czech Ann Meet Radiat Chem 10th 1970 2: 367-420

Walker DC, May R (1974) Time-resolved photobleaching of solvated electron absorption bands. Int J Radiat Phys Chem 6: 345-377

Wypych M, Kroh J, Lefik W, Plonka A (1985) Radioluminescence kinetics of frozen aqueous solutions of $MgCl_2$ and $CaCl_2$. Bull Pol Acad Sci, Chem 33: 523-528

Zaskul'nikov VM, Tolkatchev VA (1982) Kinetic law of the oxidation of ethyl radicals in alcoholic glasses taking into account structural matrix relaxation (Russ) Kin Katal 23: 1063-1070

3. KINETIC BEHAVIOUR OF WATER RADIOLYSIS PRODUCTS IN POLYCRYSTALLINE ICE

From the four possible primary species

$$H_2O^+, \quad e^-, \quad H, \quad OH \qquad (3.1)$$

of water radiolysis three have been observed as trapped in polycrystalline ice. Trapped electrons were detected by their optical absorption (Eiben and Taub 1967), trapped hydrogen atoms and trapped hydroxyl radicals were identified by ESR method (Piette et al. 1959). There seemed to be an evidence of H_2O^+ cation radicals. Attention was paid (Johnson and Moulton 1978) to a broad feature in the g = 2.08 region of the ESR spectrum observed at 4.2 K for irradiated polycrystalline ice. In fact such a feature was inevitably noticed before but dismissed as being caused by some transition-metal impurity. Johnson and Moulton demonstrated conclusively that this feature is not due to an impurity and, therefore, at 4.2 K there is in irradiated polycrystalline ice a new kind of radical in addition to OH radicals and H atoms. It is thermally reversibly radical as it is also present in ice irradiated at 77 K and cooled to 4.2 K (Johnson and Moulton 1978; Plonka et al. 1980).

In polycrystalline ice irradiated at 77 K there are trapped almost exclusively OH radicals, cf. (Kroh et al. 1961, 1962). It was, therefore, originally suggested (Johnson and Moulton 1978) that the g = 2.08 feature might be due to H_2O^+ radical cations formed in the process

$$OH + H_3O^+ \rightleftharpoons H_2O^+ + H_2O \qquad (3.2)$$

but the view was expressed (Symons 1980, 1982) that O^- was a more likely candidate

$$OH + H_2O \rightleftharpoons O + H_3O^{\prime} \; . \qquad (3.3)$$

Single crystal studies performed in our Laboratory for D_2O, in which the conversion of OD radicals upon cooling is more efficient than the conversion of OH radicals in H_2O (Johnson and Moulton 1978; Symons 1980) and line width are narrower confirmed the second process. The $g_\perp = 2.083$ and $g_\parallel = 2.002$ values coincide with those for O^- (Symons 1980) and the distinguishable groups of O^- radicals are deducible from three groups of OH (OD) radicals in I_h crystal, cf. Box (1977).

TRAPPED ELECTRONS

Trapped electrons with a visible absorption band peaking at 640 nm are formed in crystalline ice irradiated with ionizing radiation. Their half-lives around 77 K are unusually short as compared with those in

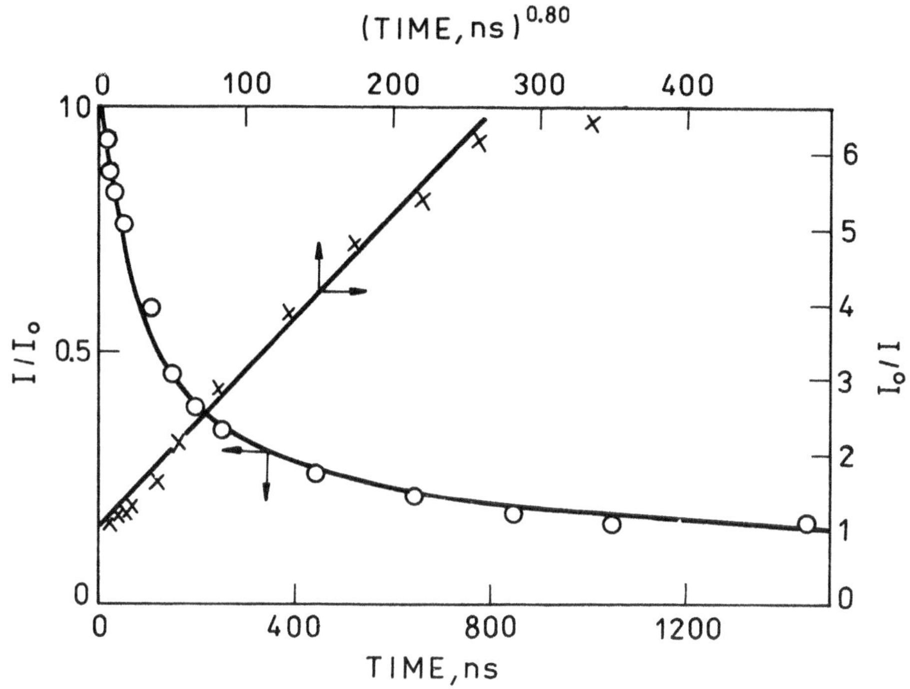

Fig. 3.1. Decay of e^-_{vis} measured at 600 nm in D_2O at 80 K. The solid line drawn through the experimental points in the coordinate system I/I_o vs. time presents the fit of eqn (3.5); the fit of eqn. (3.10) is presented in the coordinate system I_o/I vs. t^{α}

alkaline glasses discussed in Chapter 2, and there is a marked solvent isotope effect on the decay rate. Strong indications were provided (Kawabata et al. 1982) that the pair of ions H_3O^+ and e^-_{vis} are produced within their mutual Coulombing field from very highly ecited intermediates

$$H_2O^* \longrightarrow H^+ + e^- + OH \ , \qquad (3.4)$$

the ejected proton is termalized and becomes attached to H_2O forming H_3O^+ in a conductive level while e^- becomes trapped in a lattice vacancy or defect forming e^-_{vis}. OH seems to remain in their position. These electrons decay by geminate recombination with mobile H_3O^+ and thus their decay was assumed to be analogous to the geminate recombination of ion pairs in nonpolar liquids. Empirical equations for the lifetime distributions of ion pairs decaying by geminate recombination in a spur were reported by Hummel (1968) and Rzad et al. (1970). According to Hummel the fraction of ions escaping recombination at time t after formation of the ion pair is given by

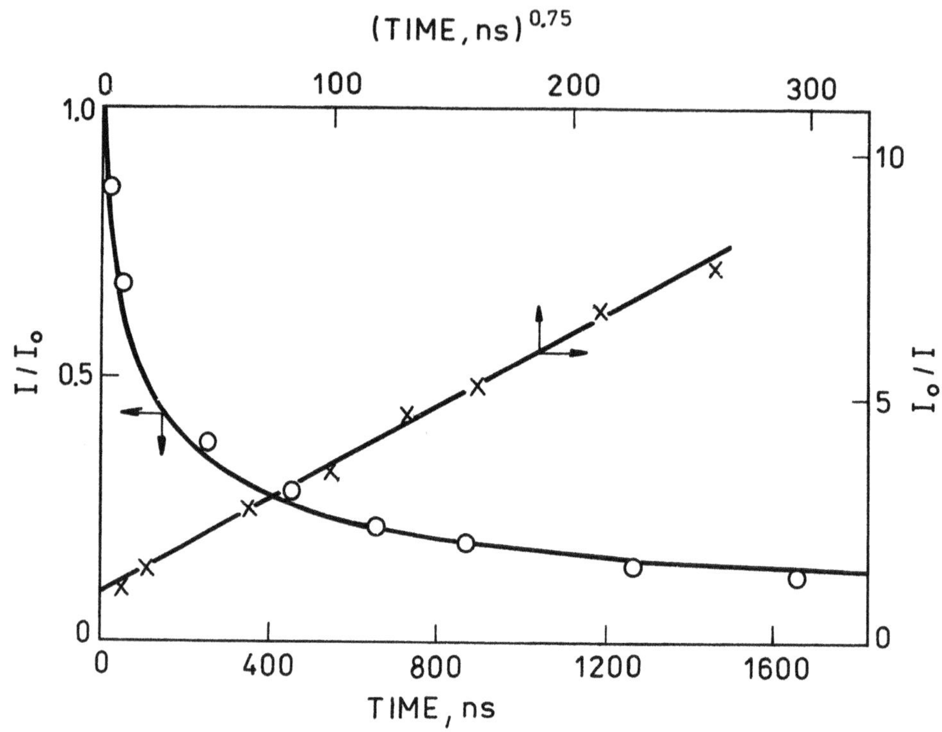

Fig. 3.2. Decay of e^-_{vis} measured at 600 nm in 80% D_2O - 20% H_2O at 80 K. The solid line drawn through the experimental points in the coordinate system I/I_0 vs. time presents the fit of eqn. (3.8); the fit of eqn. (3.10) is presented in the coordinate system I_0/I vs. t^{α}

$$G/G_0 = \mathrm{erf}\left[1/(4\lambda t)^{1/2}\right] \tag{3.5}$$

where λ is a constant and

$$\mathrm{erf}(x) = \frac{2}{\sqrt{\pi}}\int_0^x \exp(-u^2)du \tag{3.6}$$

For ions decaying during the pulse, eqn. (3.5) is corrected, cf. Rzad et al. (1970), to give the fraction remaining after the pulse

$$I = (1/\tau)\int_0^{\tau} \mathrm{erf}\left[4\lambda(t^* - t)\right]^{-1/2}dt \tag{3.7}$$

where τ denotes the pulse length and t^* is time measured from the beginning of the pulse. The integrated form of eqn. (3.7), i.e.

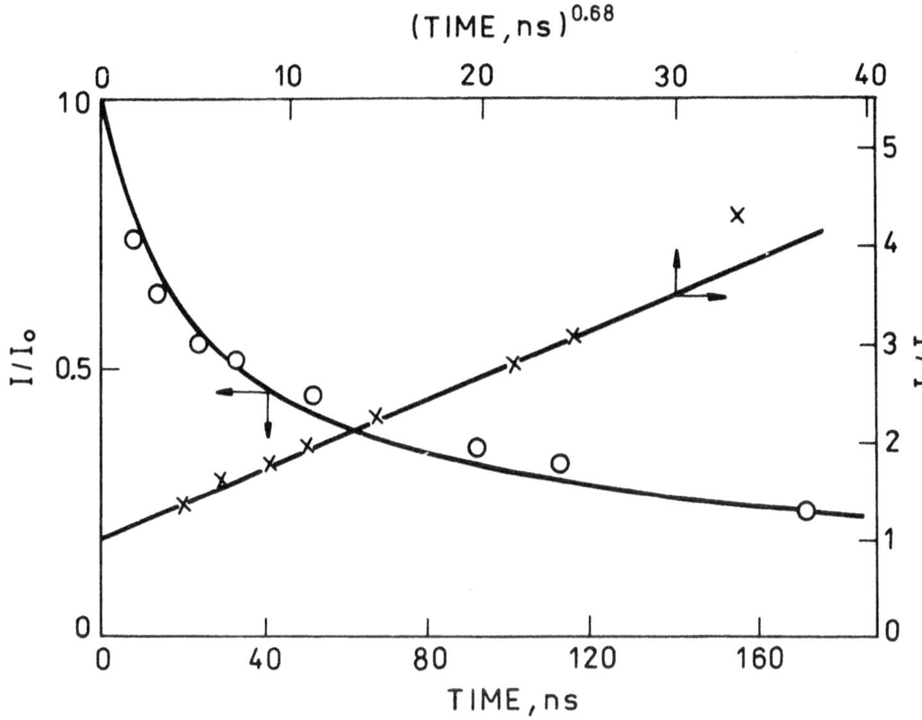

Fig. 3.3. Decay of e^-_{vis} at 600 nm in 50% D_2O - 50% H_2O at 80 K. The solid line drawn through the experimental points in the coordinate system I/I_o vs. time presents the fit of eqn. (3.8); the fit of eqn. (3.10) is presented in the coordinate system I_o/I vs. t^α

$$I = (1/2\lambda\tau)\left\{\left[\mathrm{erf}(z)\right]_{z_2}^{z_1} + (1/\sqrt{\pi})\left[(1/z)\exp(-z^2)\right]_{z_2}^{z_1} + \right.$$
$$\left. + \left[(1/2z^2)\mathrm{erf}(z)\right]_{z_2}^{z_1}\right\} \tag{3.8}$$

with

$$z_1 = (4\lambda t^*)^{-1/2}$$
$$z_2 = \left[4\lambda(t^* - \tau)\right]^{-1/2} \tag{3.9}$$

was used by Kawabata et al. (1982) to fit the experimental data. As the examples fits of eqns. (3.5) and (3.8) to the experimental values of I/I_o are shown in Figs. 3.1 - 3.5 for D_2O, 20% H_2O - 80% D_2O, 50% H_2O - 50% D_2O, 80% H_2O - 20% D_2O, and H_2O, respectively. In the same figures there are presented the fits of the second-order equal concentration kinetics with time-dependent rate constant (1.1) which in this case reads

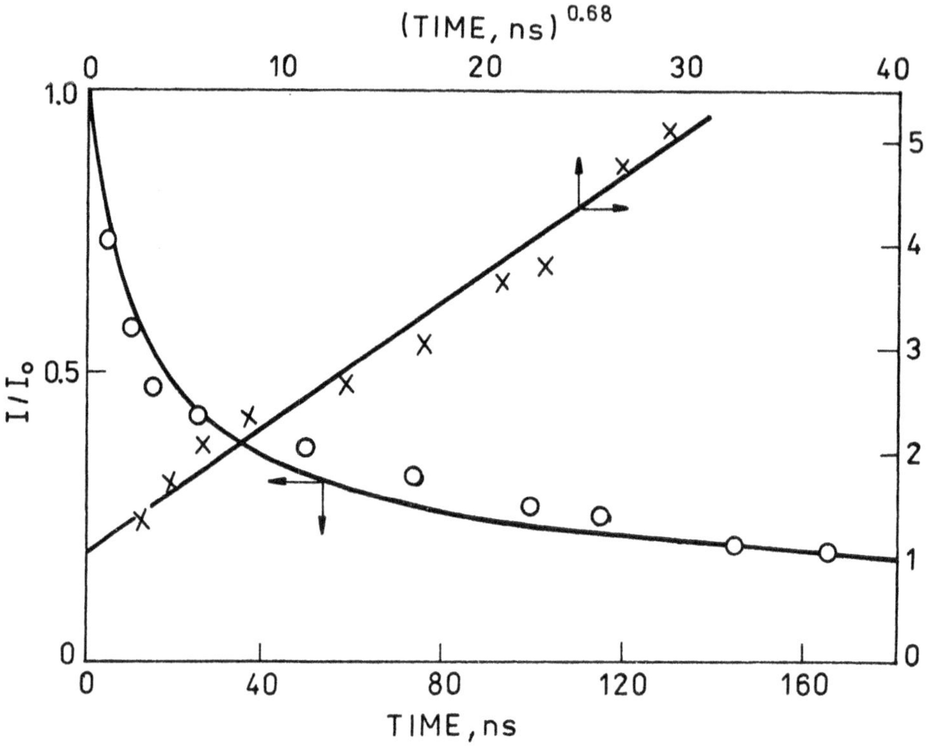

Fig. 3.4. Decay of e^-_{vis} measured at 600 nm in 20% D_2O - 80% H_2O at 80 K. The solid line drawn through the experimental points in the coordinate system I/I_o vs. time presents the fit of eqn. (3.8); the fit of eqn. (3.10) is presented in the coordinate system I_o/I vs. t^α

$$I_o/I = 1 + (I_oB/\alpha)t^\alpha \qquad (3.10)$$

and it is seen that they seem to be equally good. It is worth noting the decrease of electron half-lives from D_2O to H_2O, in the order of presentation they are equal to 117, 112, 34, 17, and 13 µs.

There is no possibility to reproduce Kawabata et al. (1982) data for temperature dependence of e^-_{vis} decay in D_2O at 6 - 100 K. It is worthwhile, however, to reproduce their calculated values of half-lives in Arrhenius plot, cf. Fig. 3.6, from which it follows that the activation energy decreases non-linearly with temperature from about 0.6 kcal/mole around 100 K to 0.02 kcal/mole or less around 15 K.

It was suggested that the decrease in the lifetime of e^-_{vis}, i.e. increase of proton mobility at higher temperatures is connected with excitation of 190 cm^{-1} phonon mode. Below 50 K these optical phonons, correlating with O...O vibrations, are thought to be completely frozen.

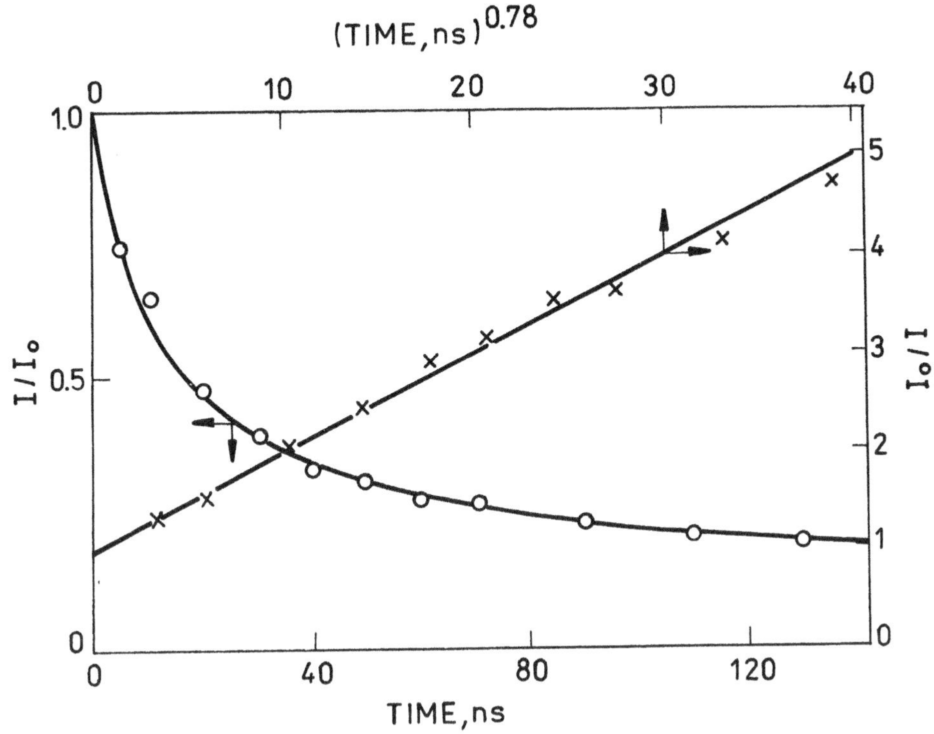

Fig. 3.5. Decay of e^-_{vis} measured at 600 nm in H_2O at 80 K. The solid line drawn through the experimental points in the coordinate system I/I_o vs. time presents the fit of eqn. (3.8); the fit of eqn. (3.10) is presented in the coordinate system I_o/I vs. t^α

HYDROGEN ATOMS

In polycrystalline ice γ-irradiated at 4 K the radiation yields of trapped hydrogen atoms and trapped hydroxyl radicals are both equal to about 0.9 radical pairs per 100 eV. When the system is warmed to 77 K all of the trapped hydrogen atoms and a small portion of trapped hydroxyl radicals disappear; the radiation yields at 77 K are zero and 0.8, respectively (Kevan 1968). Attempts to observe directly the disappearance of OH below 77 K due to reaction with H were unsuccessful (Flournoy et al. 1962) because the shape of OH spectrum was found to change sufficiently during warming to preclude the quantitative determination of relative concentration of OH at a different temperature. Now the real cause of these changes is known, cf. above, and there is no doubt that trapped hydrogen atom decay is dominated by recombination. In an effort to determine the order of decay and the activation energy, a number of rate measurements were made under isothermal conditions at several temperatures (Flournoy et al. 1962). In each series of experiments the irradiated sample was warmed from liquid helium temperature in a stepwise fashion and the concentration of trapped hydrogen atoms was followed

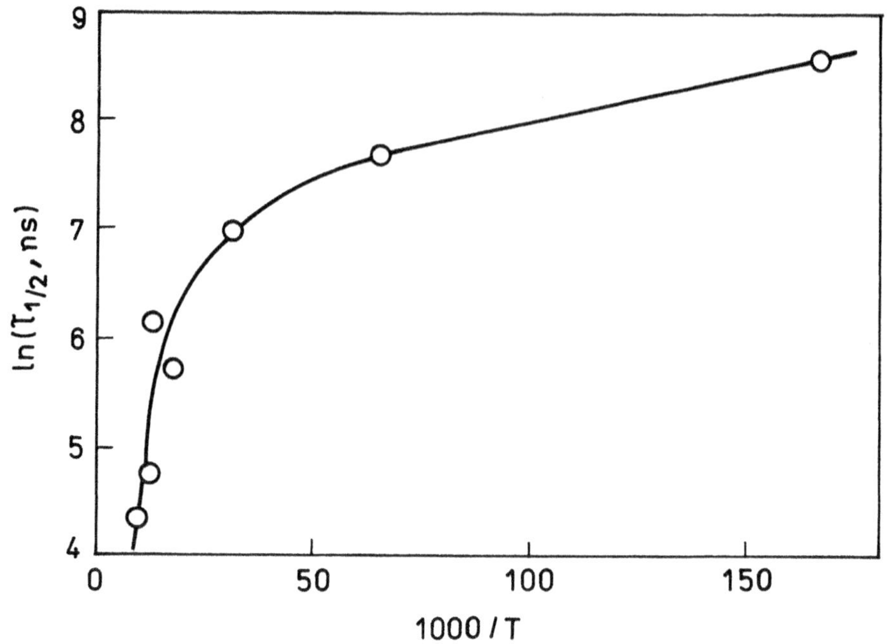

Fig. 3.6. Arrhenius plot of the half-life of e^-_{vis} in D_2O

as a function of time during several consecutive isothermal periods at succesively higher temperatures. The dependence of the isothermal rate upon the total concentration of trapped hydrogen atoms was found to be much too great to correspond with a classical rate law. At each temperature the initial rapid rate of disappearance seemed to approach a very slow asymptotic value long before all the atoms had disappeared. All these data can be fairly well fitted with eqn. (3.11)

$$[H_t]_o/[H_t] = 1 + ([H_t]_o B/\alpha)t^{\alpha} \qquad (3.11)$$

proper for the second-order equal concentration kinetics with time-dependent rate constant (1.1), cf. Fig. 3.7 showing the results of one of the two published series. With all reservation for the use of half-lives values to determine the activation energy it is instructive to plot them in Arrhenius coordinate system, cf. Fig. 3.8, to illustrate the changes of activation energy with temperature. It is seen that the apparent activation energy changes from somewhat less than 0.4 kcal/mole near 30 K to nearly 1.8 kcal/mole at about 50 K. It is interesting to note that the transformation (3.3) proceeds in the temperature region in which the abrupt changes are seen in Arrhenius plots for trapped electron decay, cf. Fig. 3.6, and trapped hydrogen atoms, cf, Fig. 3.8.

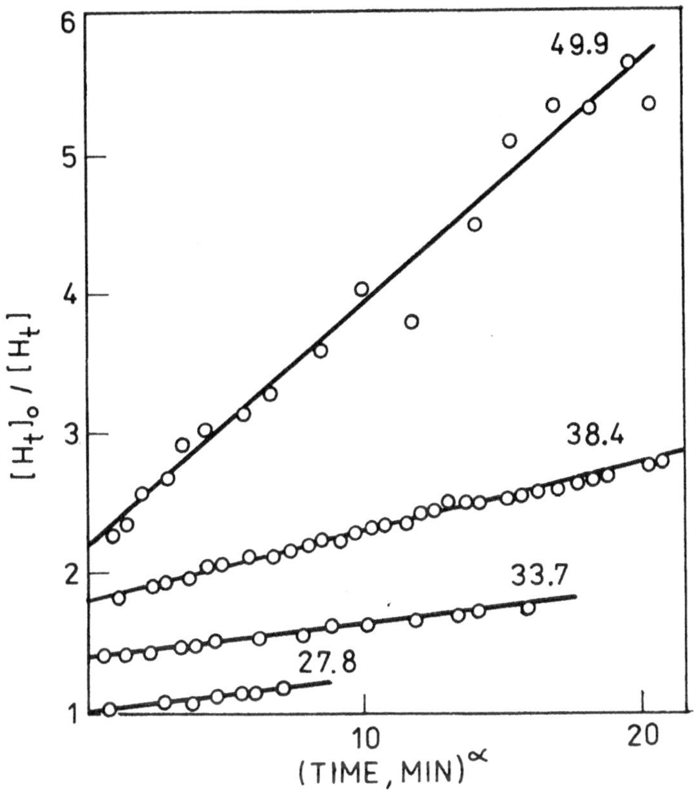

Fig. 3.7. Decay of trapped hydrogen atoms in polycrystalline ice irradiated at 4 K

HYDROXYL RADICALS

The kinetic behaviour of hydroxyl radicals trapped in ice has been open for discussion since 1960 when Siegel et al. reported the 3/2 order rate law for isothermal decay. This law, confirmed by Taub and Eiben (1968) in pulse radiolysis experiments, remained not only unexplained satisfactorily (Kevan 1968) but was also contradicted by simple first-order kinetics reported by Dibdin (1967) and Punteznis et al. (1971).

There appeared a chance to remove the discrepancies. It was noticed (Lefik 1983) that from the experimental point of view it is impossible to distinguish the 3/2 order law from the first-order kinetic equation with time-dependent rate constant of the form (1.1).

Following that idea Plonka et al. (1984) reinvestigated the hydroxyl radical decay in polycrystalline ice. It was found that at temperature below about 100 K the decay of OH radicals, cf. Fig. 3.9, can be satisfactorily described by the second-order kinetic equation with time-

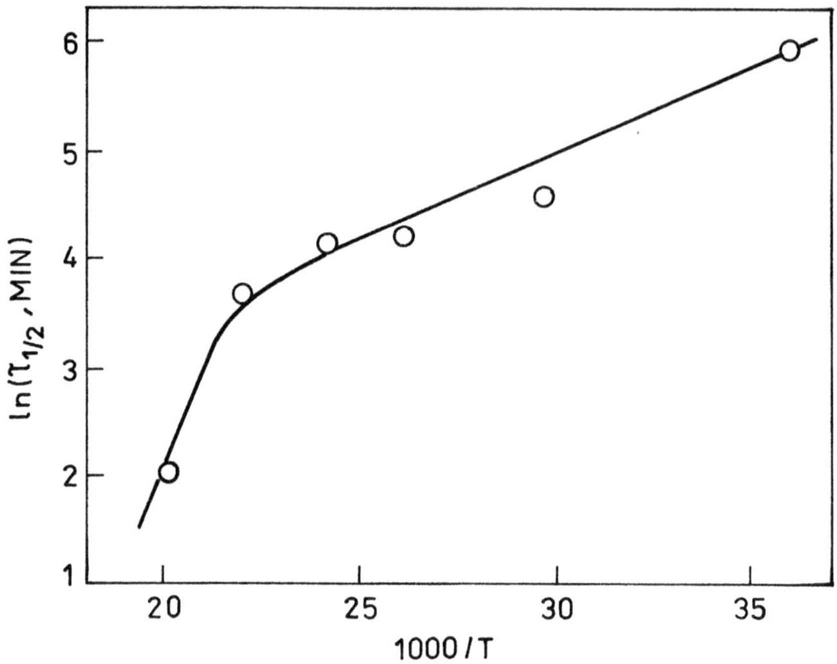

Fig. 3.8. Arrhenius plot of the half-life of H_t in polycrystalline ice

dependent rate constant

$$[OH_t]_o/[OH_t] = 1 + ([H_t]_o B/\alpha)t^{\alpha} \equiv 1 + (t/\tau_o)^{\alpha} \qquad (3.12)$$

This decay pattern can be easily reconciled with hydroxyl radical recombination

$$OH + OH \longrightarrow H_2O_2 \qquad (3.13)$$

which evidently predominates in the systems under study. Equation (3.12), however, cannot be used to describe the experimental results at temperatures above about 100 K because in these cases the second-order rate constants apparently increase with time, cf. Fig. 3.10.

Instead of equation (3.12) one can use, cf. Fig. 3.11, the first-order kinetic equation with time-dependent rate constant

$$[OH_t]/[OH_t]_o = \exp(-Bt^{\alpha}/\alpha) \equiv \exp[-(t/\tau_o)^{\alpha}] \;. \qquad (3.14)$$

Thus admitting the change of apparent reaction order in the vicinity of about 100 K one can adequately describe the recombination of OH radicals with the use of time-dependent rate constant of the form (1.1).

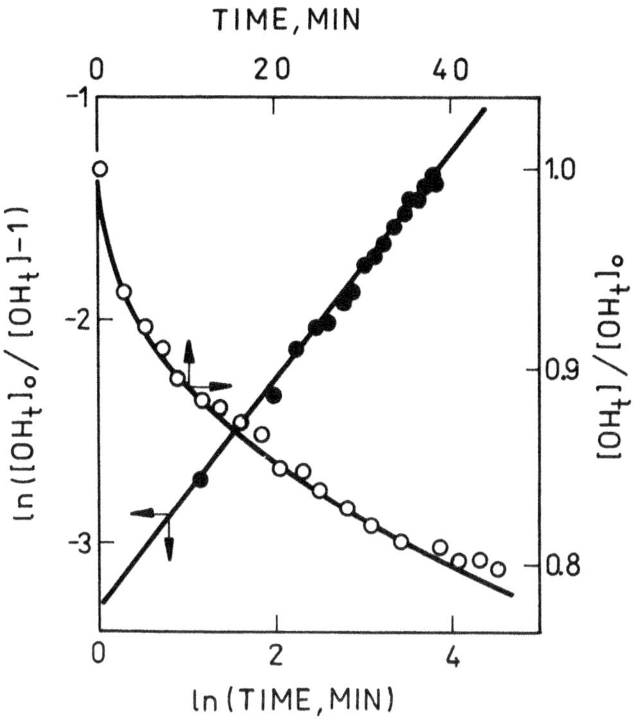

Fig. 3.9. Kinetics of OH radical decay in polycrystalline ice at 93 K presented according to eqn. (3.13); α = 0.51, τ_0 = 650 min

Smaller than unity numerical values of α obtained for the presented results indicate that there is a variety of OH radical localization sites in polycrystalline ice. It may be due to varied local properties of crystallites and surface irregularities of different size as well as to proton disorder in hydrogen bounded network of hexagonal ice (Haida et al. 1974). Even in monocrystals a number of OH trapping sites are distinguished by ESR (Box 1977).

It should be noticed that previously reported results for hydroxyl radical decay in frozen aqueous systems become consistent if one accepts the proposed kinetic treatment. Indeed, classical first-order kinetics reported by Dibdin (1967) and by Punteznis et al. (1971) correspond to equation (3.14) with α = 1, while the data of Siegel et al. (1960), cf. Fig. 3.12, are easily fitted with $\alpha < 1$. Bearing in mind that the numerical value of α depends on temperature and is affected both by irradiation time and thermal history of the sample one finds all results in fairly good agreement as far as temperatures above about 100 K are concerned. Siegel et al. (1960), however, reported the 3/2 order rate law for OH radical decay in polycrystalline ice to be valid even at temperatures much lower than 100 K. This by no means can be reconciled

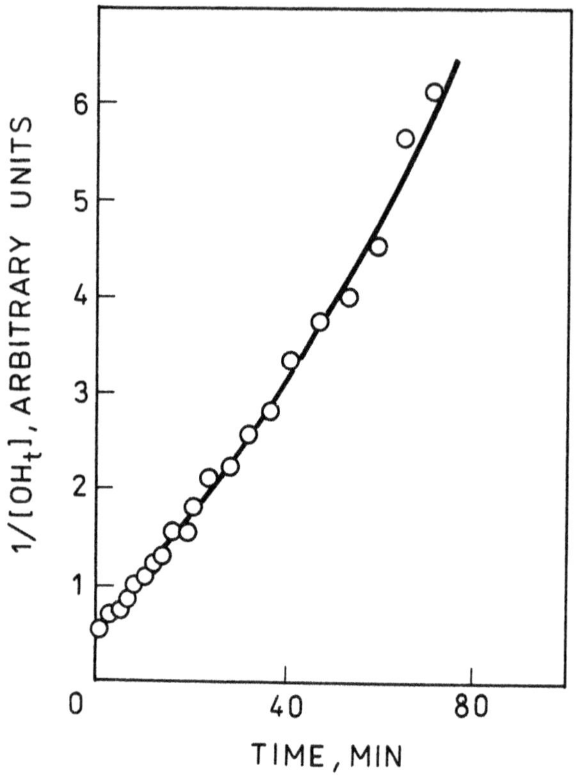

Fig. 3.10. Dependence of reciprocal concentration of OH radicals in polycrystalline ice on time at 109 K

with the second-order decay evidently observed in present studies.

There is a change of reaction order, from second to first, at a temperature region of about 100 K, in which the structural relaxation in aqueous systems is observed (Haida et al. 1974; Onsager et al. 1978; Johari and Jones 1975). This phenomenon was discussed for a number of radicals trapped in glassy matrices, cf. Chapter 2. Here it is evidenced for a polycrystalline system.

Approaching the temperature of about 100 K one observes the rapid decay of OH radical, cf. Fig. 3.13, which allows one to think of a specific temperature, which was called above the temperature of radical freeing T_f, above which the radicals are apparently mobile and recombine rapidly. Two cases are to be considered. The first one, generating the radicals below T_f and observing the recombination of radicals at the irradiation temperature or well below T_f one deals with trapped species which recombine due to their slow diffusive motion and equation (3.12) is valid. The second case, generating the radicals below T_f and observing

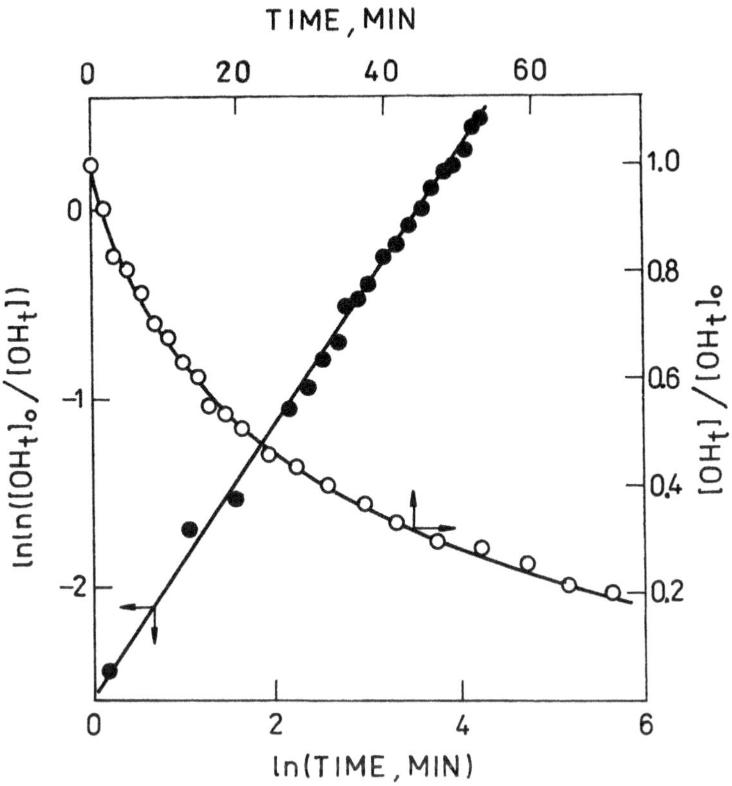

Fig. 3.11. Kinetics of OH radical decay in polycrystalline ice at 109 K presented according to eqn. (3.14); α = 0.72, τ_0 = 36.5 min

the recombination of radicals at or above T_f, one deals with rapidly reacting mobile species but the rate determining step consists in radical freeing. Because of this the first-order decay, adequately described by eqn. (3.14) is observed. There are reasons to believe that the distinction between trapped and mobile species in a given matrix is rather quantitative than qualitative and following the recombination at the irradiation temperature, below or above T_f, one always observes the second-order kinetics.

STRUCTURAL RELAXATION IN POLYCRYSTALLINE ICE

It is known that during thermal annealing of polycrystalline ice irradiated at the liquid nitrogen temperature to the melting point, the system passes through three widely separated relaxation regions (Johari and Jones 1975), at 115±5, 160±30, and 230±30 K, in which, due to microscopic displacement of solvent molecules, the trapped species could be made mobile and/or reactive.

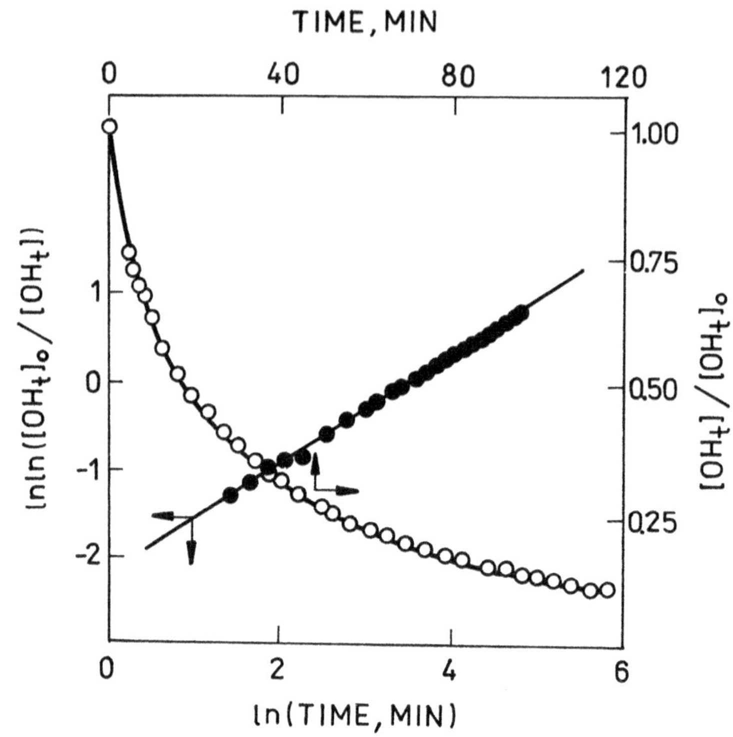

Fig 3.12. Data of Siegel et al. (1960) for decay of OH radicals in polycrystalline ice at 100 K presented according to eqn. (3.14); α = 0.63, τ_o = 33.0 min

In the first relaxation region the OH radicals decay rapidly due to recombination, cf. eqn. (3.13), followed to some extent by

$$OH + H_2O_2 \longrightarrow HO_2 + H_2O \ . \qquad (3.15)$$

Thus formed HO_2 radicals decay in the third relaxation region due to dismutation

$$HO_2 + HO_2 \longrightarrow H_2O_2 + O_2 \ . \qquad (3.16)$$

The existence of distinct relaxation regions in ice in which the trapped species are selectively made reactive is of great practical interest. As an example, in the second relaxation region in which in pure polycrystalline ice the hydroperoxy radicals decay completely we have observed (Plonka et al. 1978) in the presence of native superoxide dismutase the loose complex, [E-Cu^{2+}...HO_2]. The reaction of HO_2 dismutation starts in the third relaxation region. The decay of the ESR signal of superoxide radicals is not accompanied by immediate changes of the oxidation state of copper in the enzyme. There is marked delay of the

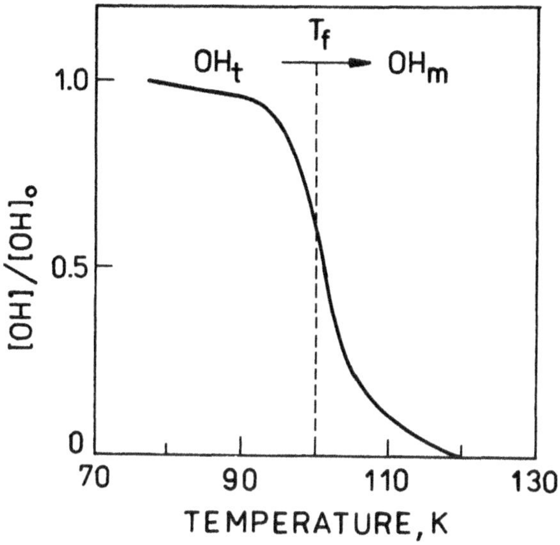

Fig. 3.13. Thermal stability of hydroxyl radicals in polycrystalline ice γ-irradiated at 77 K

changes of the oxidation state of copper, proceeding at about 270 K, which indicates that dismutation of superoxide radicals proceeds without changes of the oxidation state of copper in superoxide dismutase and contradicts the reaction scheme involving alternate use of the two active centres present in superoxide dismutase (Plonka et al. 1978, 1980). The delayed changes in the oxidation state of copper in enzyme are due to the reactions with products of dismutation reaction (Plonka et al. 1980) leading to partial reduction of native superoxide dismutase or to partial oxidation of reduced superoxide dismutase, respectively.

There seems to be at least one relaxation region below 77 K as indicated not only by the abrupt changes of the activation energy for electron, cf. Fig. 3.6, and hydrogen atom decay, cf. Fig. 3.8, by also by transformation (3.3). The O^- radicals are observable up to about 24 K (Johnson and Moulton 1975).

SOME OTHER POLYCRYSTALLINE SYSTEMS

Radicals produced by γ-irradiation of polycrystalline hydrocarbons undergo stepwise decay as the temperature is raised to successively higher temperatures (Ermolaev et al. 1962a, 1962b; Mikhailov et al. 1965). The decay patterns are similar to those discussed above for trapped hydrogen atoms. As an example the data of Wilkey et al. (1977) for cracked crystals of n-hexane are reproduced in the insert to Fig. 3.14. No measurable decay of the radicals produced by γ-irradiation of n-hexane took place in 1 week at the irradiation temperature of 77 K. As the temperature was raised the observable decay began at about 0.8 of

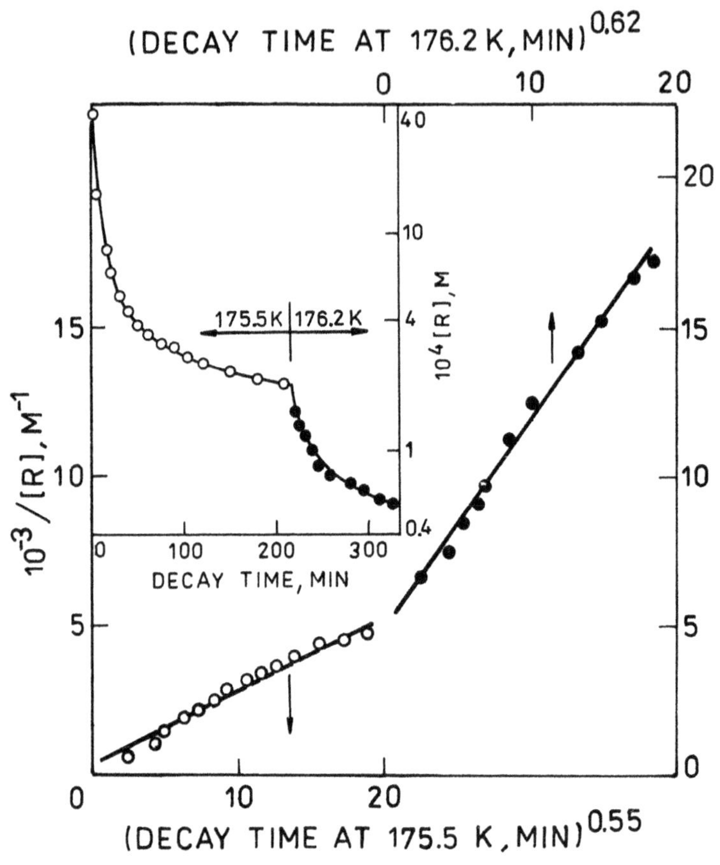

Fig. 3.14. Decay of radicals in severely cracked crystals of n-hexane following γ-irradiation at 77 K

the melting point. At each temperature there was an initial rapid decay followed by a decrease in rate to a very low value. Neither during the rapid decay nor the slow decay did the decay follow a linear second order plot. It can be well fitted with the second-order kinetic equation with time-dependent rate constant (1.1), cf. Fig. 3.14. In some cases one will have to admit the change of reaction order.

REFERENCES

Bensasson R, Durup M, Dworkin A, Magat M, Marx R, Szwarc H (1963) Role of molecular motions in reactions of radicals produced by irradiation and trapped solid matrixes. Discuss Faraday Soc 36: 177-185

Box HC (1977) Radiation Effects: ESR and ENDOR analysis. Academic press, New York San Francisco London

Dibdin GH (1967) ESR of γ-irradiated single crystals of ice at 77° K.

Identification of the hydroxyl radical and its trapping site. Trans Faraday Soc 63: 2098-2111
Eiben K, Taub IA (1967) Solvated electron spectrum in irradiated ice. Nature 216: 782-783
Ermolaev VK, Molin YuN, Buben NYa (1962) Recombination of radicals in organic solids. I. Investigation by melting (Russ). Kinet Katal 3: 58-64
Ermolaev VK, Molin YuN, Buben NYa (1962) Recombination of radicals in organic solids. II. Kinetic regularities (Russ). Kinet Katal 3: 314-321
Flournoy JM, Baum LH, Siegel S (1962) Disappearance of trapped hydrogen atoms in gamma-irradiated ice. J Chem Phys 36: 2229-2230
Haida O, Matsuo T, Suga H, Seki S (1974) Calorimetric study of the glassy state. X. Enthalpy relaxation at the glass transition temperature of hexagonal ice. J Chem Thermodyn 6: 815-825
Hummel A (1968) Lifetime of charged species in irradiated dielectric liquids. J Chem Phys 49: 4840-4844
Johari GP, Jones SJ (1975) Study of the low-temperature "transition" in ice I_h by thermally stimulated depolarization measurements. J Chem Phys 62: 4213-4223
Johnson JE, Moulton GC (1978) ESR study of ice irradiated at 4.2 K, a thermally reversible radical. J Chem Phys 69: 3108-3111
Kawabata K, Nagata Y, Okabe S, Kimura N, Tsumori K, Kawanishi M, Buxton GV, Salmon GA (1982) Fast decay of the visible band electron in e-irradiated crystalline ice at low temperature: The isotope effects and the role of a mobile proton in the decay. J Chem Phys 77: 3884-3892
Kevan L (1968) Radiation chemistry of frozen aqueous solutions. In: Stein G (ed) Radiation Chemistry of aqueous systems. The Weizmann Science Press of Israel, Jerusalem, pp 21-71
Kroh J, Green BC, Spinks JWT (1961) Electron paramagnetic resonance (EPR) studies on the production of free radicals in hydrogen peroxide at liquid nitrogen temperature. J Am Chem Soc 83: 2201-2202
Kroh J, Green BC, Spinks JWT (1962) Electron paramagnetic resonance studies on free radicals produced by T β-particles in frozen H_2O and D_2O media at liquid nitrogen temperature. Can J Chem 40: 413-425
Lefik W (1983) Ph D Thesis, Technical University of Lodz, Poland
Mikhailov AI, Lebedev YaS, Buben NYa (1965) Stepwise recombination of free radicals in irradiated organic compounds. II. Formal kinetic model and a method for evaluation of kinetic constants. (Russ). Kinet Katal 6: 48-55
Onsager L, Staebler DL, Mascarenhas S (1978) Electrical effects during condensation and phase transition of ice. J Chem Phys 68: 3823-3828
Piette LH, Rempel RC, Weaver HE, Flournoy JM (1959) EPR studies of electron irradiated ice and solid hydrogen. J Chem Phys 30: 1623-1624
Plonka A, Bednarek J, Bogus W (1980) ESR investigations of thermally reversible species in γ-irradiated polycrystalline ice. Bull Acad Polon Sci, Sér Sci Chim 28: 133-137
Plonka A, Metodiewa D, Gasyna Z (1980) ESR studies on oxidation state changes of copper in superoxide dismutase during reactions with water radiolysis products at cryogenic temperatures. Biochim Biophys Acta 612: 299-304
Plonka A, Metodiewa D, Jeziorowska A (1978) ESR evidence of paramagnetic intermediate formation from water radiolysis products and SOD. Biochem

Biophys Res Comm 81: 1344-1351
Plonka A, Szajdzinska-Pietek E, Kroh J (1984) Decay kinetics of hydroxyl radicals in frozen aqueous systems. Radiat Phys Chem 23: 583-587
Punteznis SA, Ershov BG, Pikaev AK (1971) Effect of impurities on the accumulation and loss of hydroxyl radicals in crystalline ice irradiated with γ-rays at 77° K (Russ). Dokl Akad Nauk SSSR 196: 641-644
Rzad SJ, Infelta PP, Warman JM, Schuler RH (1970) Kinetics of electron scavenging and ion recombination in the radiolysis of hydrocarbon solution. J Chem Phys 52: 3971-3983
Siegel S, Baum LH, Skolnik S, Flournoy JM (1960) Observations of the thermal behavior of radicals in gamma-irradiated ice. J Chem Phys 32: 1249-1256
Symons MCR (1980) Conversion of OH radicals into O^- ions on cooling: a radiation - ESR study. J Chem Soc, Chem Commun 675-676
Symons MCR (1982) Electron spin resonance studies of OH and O^- radicals in irradiated ice. J Chem Soc, Faraday Trans I 78: 1953-1959
Taub IA, Eiben K (1968) Transient solvated electron, hydroxyl and hydroperoxy radicals in pulse-irradiated crystalline ice. J Chem Phys 49: 2499-2513
Wilkey DD, Fenrick HW, Willard JE (1977) γ-irradiated hydrocarbon crystals. Yields, decay, and photoreactions of radicals: carbanion formation. J Phys Chem 81: 220-228

4. DECAY OF RADICALS IN POLYMER MATRICES

It was rationalized (Emanuel 1981) that to describe the kinetic processes involving radicals in polymer matrices one has to use an integro-differential equation which for the case of simple monomolecular reactions reads

$$-d[R]/dt = \int_0^\infty kR(k)dk \qquad (4.1)$$

where R(k) denotes the distribution of radicals according to their reactivity expressed in terms of reaction rate constant k.

At temperatures below the glass transition of the polymer matrix, the kinetic processes described by eqn. (4.1) can be assumed to proceed in many small regions with constant local properties. Then the solution of eqn. (4.1) is given by a Fredholm integral equation of the first kind

$$[R] = \int_0^\infty R(k)\exp(-kt)dk \, . \qquad (4.2)$$

Thus the experimental kinetic curve reflects, in terms of Laplace transform, the initial distribution of R(k), cf. Appendix B.

It was shown above that in disordered matrices, such as low-temperature glasses or polycrystalline solids, the solution of Fredholm equation (4.2) can be satisfactorily approximated by

$$[R] = [R]_0\exp(-Bt^\alpha/\alpha) \, , \qquad (4.3)$$

Equation (4.3) is, cf. Appendix A, the first-order kinetic equation with the time-dependent rate constant of the form

$$k(t) = Bt^{\alpha-1} \, , \quad 0 < \alpha \leqslant 1 \, . \qquad (4.4)$$

This particular form (4.4) was shown to also describe adequately the decay patterns of radicals in polymer matrices (Plonka 1983).

ALLYL RADICALS IN POLYETHYLENE

The results of Auerbach and Sanders (1969), recently discussed by Hwang and Cheng (1981), on allyl radical recombination in irradiated polyethylene (Marlex 50) are suitable for direct illustration of the applicability of time-dependent rate constant (4.4) to describe the kinetic patterns of radical decay of polymers. These radicals recombine according to the second-order rate equation with the observed rate constants reproduced in Fig. 4.1. Through the experimental points there are drawn lines corresponding to eqn. (4.4) with the constants B and α (cf. Table 4.1) obtained by linearization of the experimental results in coordinate system ln k(t) vs. ln t. From the comparison presented it is evident that eqn. (4.4) describes quite well the time dependence of the observed reaction rate at least as well as the much more elaborate treatment of Hwang and Cheng (1981) taking into account cage effects.

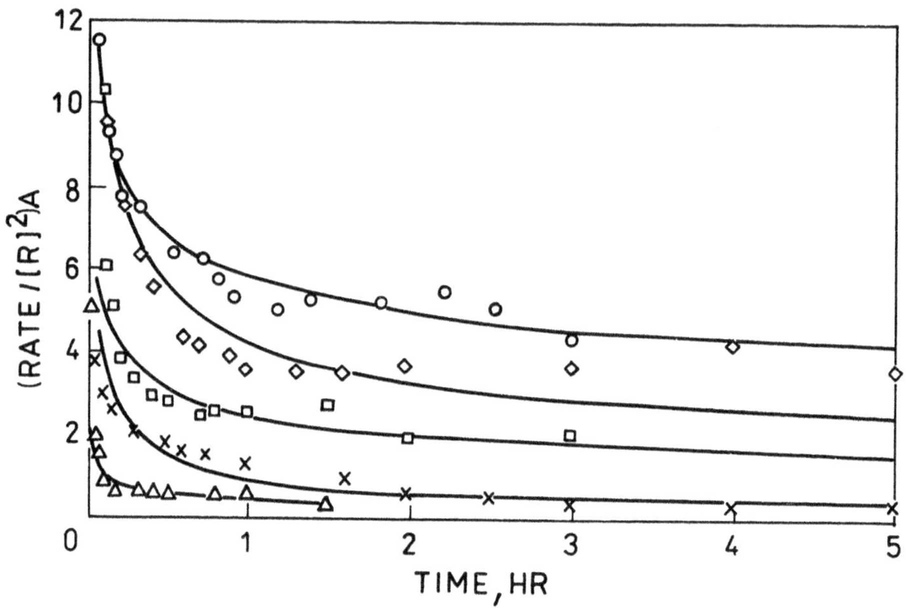

Fig. 4.1. Comparison with eqn. (4.4) of experimental results of Auerbach and Sanders (1969) on time dependence of observed reaction rate for allyl radical decay in polyethylene. The ordinate unit, according to Butyagin's (1972) correction for sample 1, is equivalent to $2.9 \cdot 10^{19}$ cm^3/radical s; the ordinate factor A and the constants B and α obtained by fitting curves 1 to 5 (from top to bottom) are given in Table 4.1

To present a more direct comparison with the experiment the data of Patel et al. (1978) on allyl radical decay in hydrogenated polyethylene are reproduced in Fig. 4.2 as $(1/[R] - 1/[R]_0)$ vs. t and $\ln(1/[R] - 1/[R]_0$ vs. ln t. In the latter plot the second-order kinetic euqtion with time-dependent rate constant (4.4)

$$[R]^{-1} - [R]_0^{-1} = Bt^{\alpha}/\alpha \qquad (4.5)$$

is linearized. The experimental points fit this relation perfectly, cf. Fig. 4.2. In this case neither Waite's (1958) treatment nor that of Dole et al. (1975), assuming two simultaneous second-order reactions to occur in different zones, is adequate (Patel et al. 1978).

Table 4.1. Experimental conditions and rate parameters calculated from results of Auerbach and Sanders (1969)

Sample	A[a]	T K	Dose Mrad	$B \cdot 10^{20}$ cm^3/radical $s^{-\alpha}$	α
1	10^3	343	0.057	0.40	0.79
2	10^4	343	1.54	0.05	0.64
3	10^3	391	28.5	0.26	0.69
4	10^4	343	27.4	0.03	0.38
5	10^2	398	28.5	0.58	0.61

[a] A is the ordinate factor in Fig. 4.1.

Table 4.2. Experimental conditions and rate parameters calculated from results of Patel et al. (1978)

Polyethylene	Amorphous fraction volume content	T K	Dose Mrad	$B \cdot 10^{20}$ cm^3/radical $s^{-\alpha}$	α
Hydrogenated	0.338	353	39.8	2.01	0.61
Marlex film	0.272	353	39.8	0.92	0.75
		373	39.8	2.44	0.81
Extended-chain	0.036	393	7.6	1.36	0.82
		393	18.7	0.90	0.60
		373	39.8	0.42	0.55
		393	39.8	0.53	0.61

The same is true for all other results presented by Patel et al. (1978) (cf. Table 4.2 for the calculated rate-constant parameters) with but one exception: the very rapid decay of allyl radicals in extended chain polyethylene at 408 K, cf. Fig. 4.3. It is hard to accept the authors' (Patel et al. 1978) statement that the decay is almost exactly pure second-order with a decay constant that increases slightly with time. These data could not be fitted to relation (4.5) with $\alpha \leq 1$. One has to accept, as in other cases discussed above, the change of reaction order for recombination and to fit the experimental data by a first-order equation with a time-dependent rate constant, eqn. (4.3); which yields a straight line in a plot ln $\ln([R]_0/[R])$ vs. ln t. The perfect fit of experimental data to eqn. (4.3) with B = 0.15 $min^{-\alpha}$ and α = 0.82 is presented in Fig. 4.3. The same is probably true for allyl radical decay in Marlex bulk film at 383 K. The data, however, are too scattered to be discussed. The explanation of the change in reaction order is that at temperature close to that of the matrix structural relaxation the kinetic processes are dominated by the first-order detrapping of reactive species.

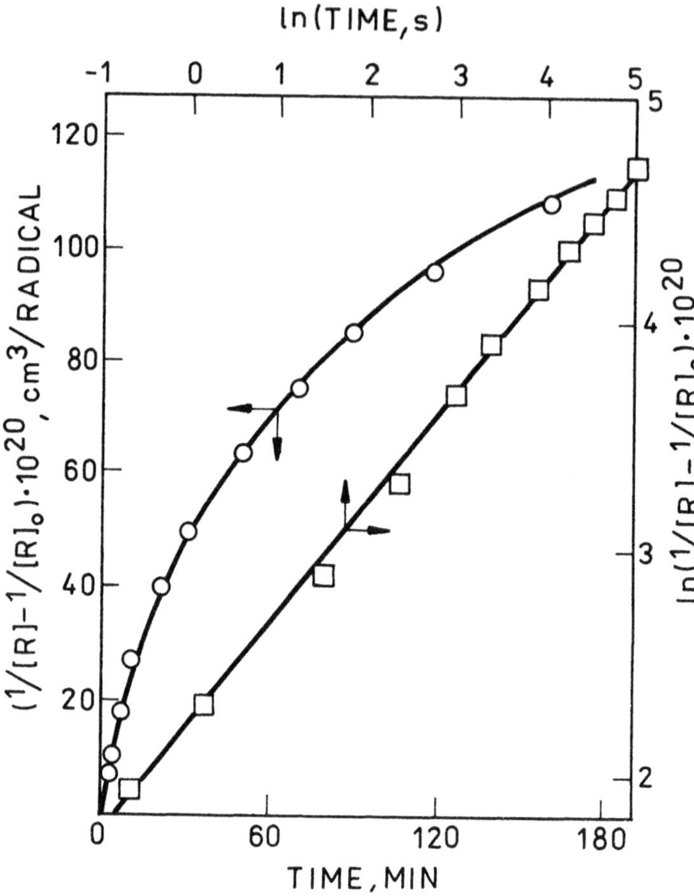

Fig. 4.2. Comparison with eqn. (4.5) of experimental results of Patel et al. (1978) on allyl radical decay in hydrogenated polyethylene at 353 K after irradiation to 39.8 Mrad at 77 K. The constants B and α obtained by curve fitting are included in Table 4.2

Basheer and Dole (1983) have used eqn. (4.5) to describe the decay of allyl radicals in linear low-density polyethylene (LLDPE) with the results presented in Table 4.3. The perfect fit at 353 K was emphasized, cf. Fig. 4.4. At lower temperatures, such as 333 K, some deviations were observed which were probably due to erroneous estimation of $[R]_0$ by extrapolation from the plot $\ln[R]$ vs. $t^{1/2}$.

As has been noted, the kinetic curves reflect the initial distribution of radicals, i.e. the distribution after irradiation. If one can neglect the effects of irradiation on matrix properties, the prolonged irradiation eliminates the more reactive species and therefore leads to narrower distributions: i.e. B decreases and α increases with dose at

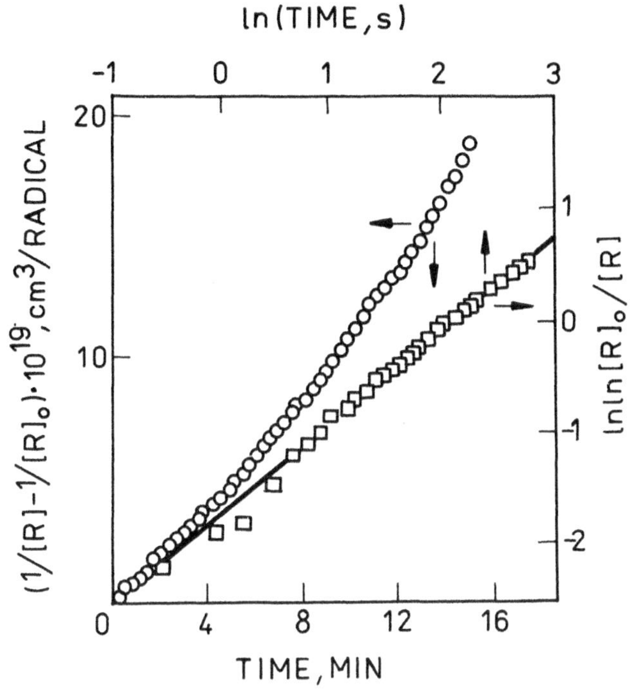

Fig. 4.3. Comparison of experimental results of Patel et al. (1978) on allyl radical decay in extended-chain polyethylene at 408 K after irradiation to 39.8 Mrad at 77 K with eqn. (4.3)

Table 4.3. Experimental conditions and rate parameters for allyl radical decay in LLDPE, Basheer and Dole (1983)

Polyethylene	T,K	$10^{-3}[R]_o$ [a]	α	10^5B [b]
QLLDPE	343	7.35	0.60	1.33
ALLDPE	343	7.19	0.58	1.03
QLLDPE	353	7.48	0.70	2.12
ALLDPE	353	7.63	0.63	1.72

Q - quenched, A - annealed sample
a - relative units
b - $min^{-\alpha}$

a constant dose rate. In polyethylene matrices, however, with increased dose more crosslinks are produced in amorphous regions; this leads to a decrease in the mobility of the chain segments containing the allyl radicals and a decrease of decay rate with dose is accompanied by a less uniform distribution (cf. Table 4.1). An increase of dose from 0.057 to 27.4 Mrad is followed by a decrease of B from $0.40 \cdot 10^{-20}$ cm^3/radical s

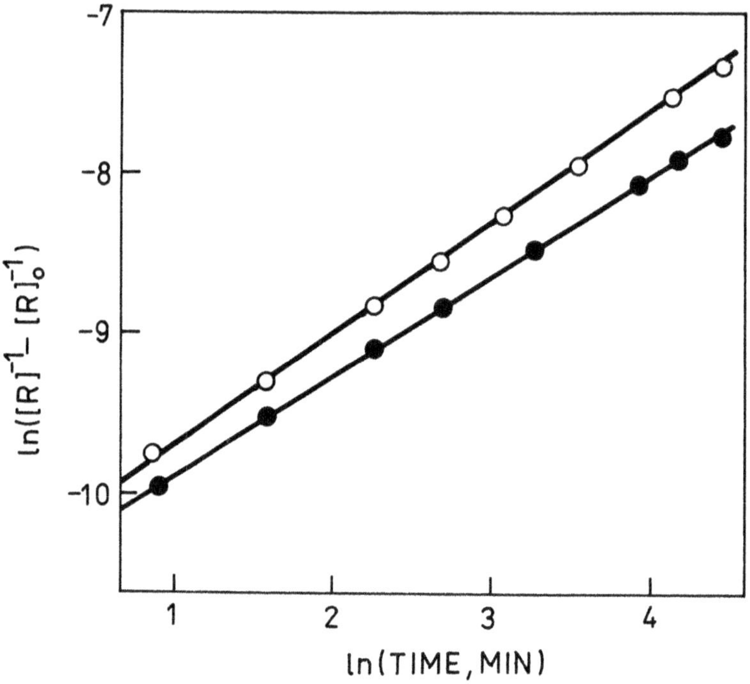

Fig. 4.4. Test of eqn. (4.5) for allyl radical decay at 353 K in the quenched (O) and annealed (●) samples of LLDPE

to $0.03 \cdot 10^{-20}$ cm^3/radical $s^{-\alpha}$ and a decrease of α from 0.79 to 0.38.

The higher fraction of amorphous content in the polyethylene matrix, the higher is the decay rate (cf. Table 4.2). At 353 K the decay is faster in hydrogenated polyethylene than in Marlex film despite the considerably lower initial concentration of allyl radicals in the former ($0.707 \cdot 10^{18}$ spins/cm^3) as compared with the latter ($1.995 \cdot 10^{18}$ spins/cm^3). At 373 K the decay in extended-chain polyethylene is considerably smaller than in Marlex film, the initial concentrations being (Patel et al. 1978) $2.10 \cdot 10^{18}$ and $1.69 \cdot 10^{18}$ spins/cm^3, respectively.

An increase in temperature, promoting the relaxation and diffusion processes in the matrix, changes the distribution of radicals, as displayed by the increasing magnitude of α , and increases their intrinsic decay rate (cf. Tables 4.1 and 4.2). Assuming an Arrhenius type temperature dependence for B, one can estimate from data for similar radical distributions the activation energy of allyl radical decay. For Marlex film it is 12.8 kcal/mole as compared to 10.6 - 14.3 kcal/mole according to other estimates from the same experimental data (Patel et al. 1978). All these findings were fully supported by the results obtained for γ-irradiated poly(methyl methacrylate)(Plonka and Pietrucha 1983).

PROPAGATING RADICALS IN POLY(METHYL METHACRYLATE)

Fig. 4.5 illustrates the thermal stability of the propagating radicals:

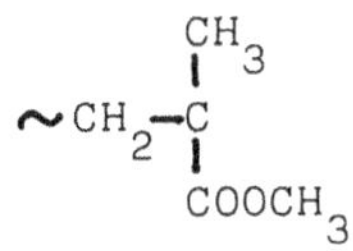

with well known nine line ESR spectrum (Szöcs et al. 1977) in PMMA γ-irradiated at 298 K.

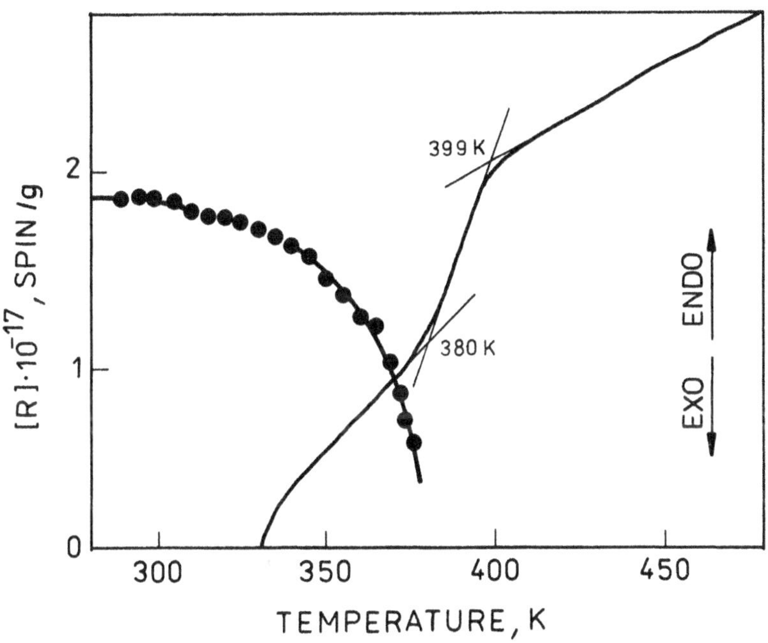

Fig. 4.5. Decay of propagating radicals in PMMA γ-irradiated at 298 K during thermal annealing for 10 min at succesively higher temperatures (●). The DSC curve showing the glass transition region is depicted

The rapid decay of these radicals starts just below the temperature region of glass transition at the onset of α-relaxation process. In the α relaxation region, beginning for conventional PMMA at about 353 K, chain backbone motions prevail (ibid.). These are the motions which enable the transport of free valences to the neighbouring places until recombination of two close radicals occurs. It is to be believed that chain backbone motions are not completely suppressed even far below the glass transition region to account for the slow decay of propagating radicals at these temperatures. Indeed, it was stressed above that molecular reorientation coupled with a certain specific defect, as a trapped radical surely is, may be very fast even at low temperatures.

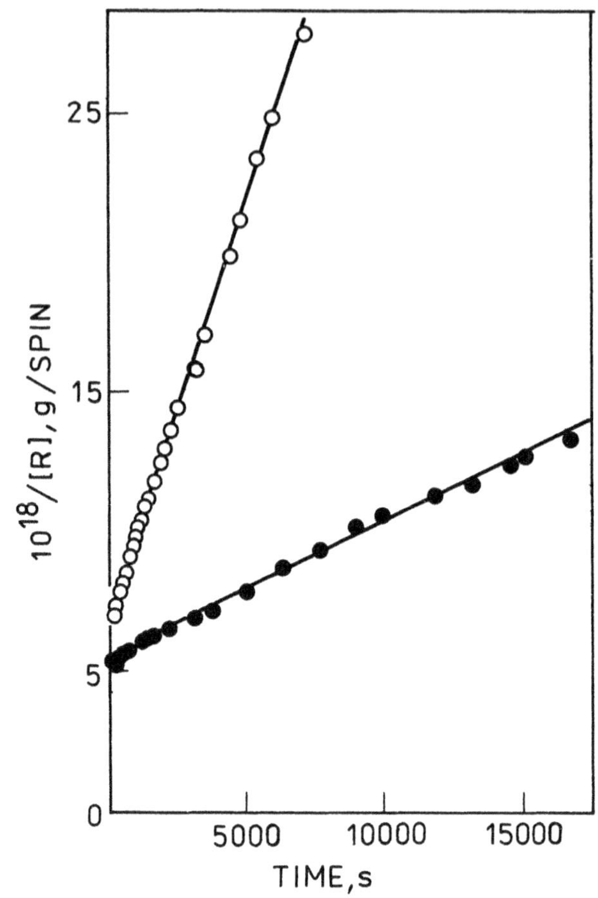

Fig. 4.6. Decay of propagating radicals at 346 (●) and 353 K (○) in PMMA γ-irradiated at 298 K

Following the isothermal decay of the propagating radicals at temperatures well below the glass transition region, say 346 K, cf. Fig. 4.6, one finds that to reconcile the observed decay with the second-order kinetics one has to disregard the initial period of more rapid decay. This initial period vanishes with the increase of temperature and at 353 K, cf. Fig. 4.6, one can claim that the experimental points do fit the second-order kinetics. However, further increase of temperature, cf. Fig. 4.7, exhibits new deviations from the second-order kinetics and one would rather try to reconcile the observed decay with the first-order kinetics at least in the initial period.

The same kind of deviations from the second-order kinetics one observes for the samples γ-irradiated at 77 K, cf. Fig. 4.8. However, due to much lower irradiation temperature they are evident at a somewhat lower

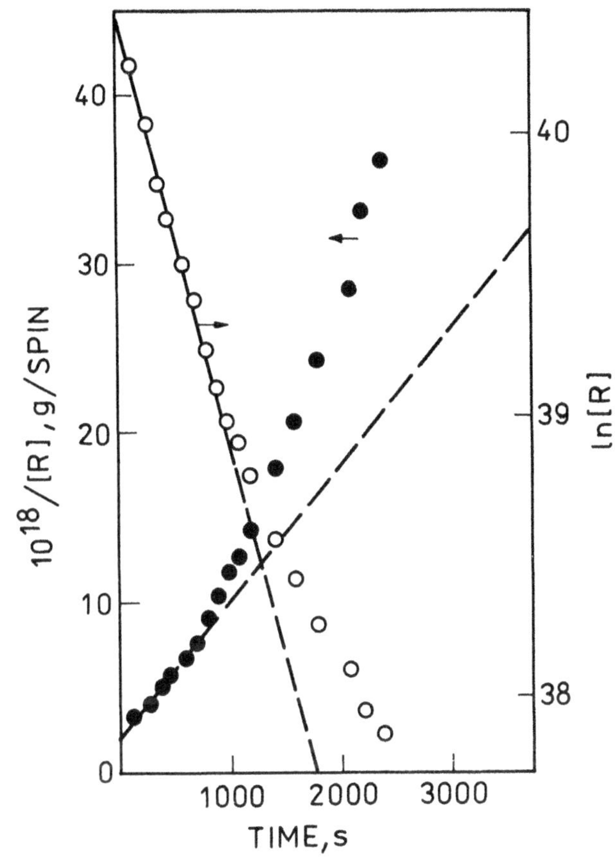

Fig. 4.7. Decay of propagating radicals at 365 K in PMMA γ-irradiated at 298 K

temperature, cf. Figs. 4.7 and 4.8.

To stress the effect of irradiation temperature on the decay kinetics there are presented in Fig. 4.9 the results for propagating radical decay at 353 K in PMMA γ-irradiated at 343, 353, and 360 K. Not only the increase of decay rate with decrease of irradiated temperature but also the change of decay patterns is evident.

Second-order kinetic equation with time-dependent rate constant, eqn. (4.5), is perfectly fitted by the experimental points from Fig. 4.6, 346 K, and Fig. 4.9, 353 and 360 K, cf. Fig. 4.10 presenting the experimental data in the coordinate system $\ln(1/[R] - 1/[R]_o)$ vs. $\ln t$. To stress the adequacy of the proposed kinetic treatment there are presented in Figs. 4.11 and 4.12 in the same coordinate systems some representative examples of the published kinetic data for similar systems well below the glass transition region.

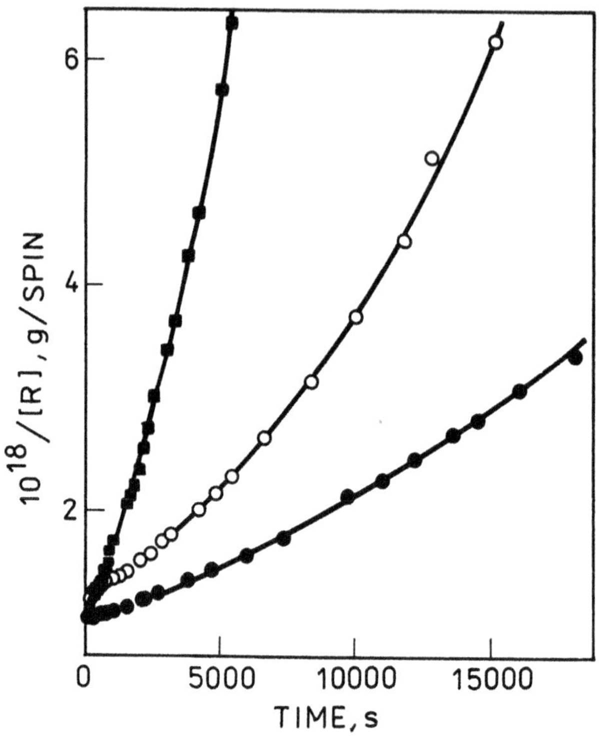

Fig. 4.8. Decay of propagating radicals at 345 (●), 356 (○), and 363 K (■) in PMMA γ-irradiated at 77 K

At temperature close to that of glass transition region there is the change of reaction order for bimolecular reaction and the experimental data are expected to fit the first-order kinetic equation with time-dependent rate constant, eqn. (4.3), i.e. to yield the straight line in the coordinate system ln $\ln([R]_o/[R])$ vs. ln t. This is the case, cf. Fig. 4.13, which illustrates the fit of experimental data from Figs. 4.7 and 4.8 to relation (4.3). The numerical values of the rate para meters, B and α , calculated from the discussed data are presented in Table 4.4.

The constant B could be taken as the intrinsic rate constant for the process under discussion and is seen to depend not only on the temperature at which the process is followed but also on the irradiation temperature. Radicals trapped in matrix irradiated at a temperature higher than that of observation are more stable, and radicals trapped in matrix irradiated at a temperature lower than that of observation are less stable than those formed at the temperature of observation. This phenomenon was seen above for trapped hydrogen atoms but more experimental data are needed to follow quantitatively the above presented relations.

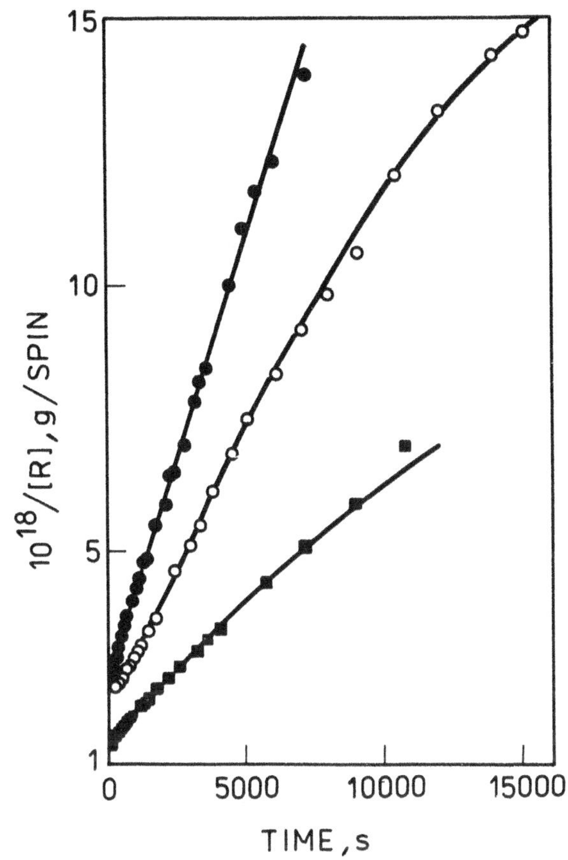

Fig. 4.9. Decay of propagating radicals at 353 K in PMMA γ-irradiated at 343 (●), 353 (○), and 360 K (■)

To determine the activation energy one has to use the temperature dependence of B for constant numerical value of α. For PMMA γ-irradiated at room temperature, taking the recalculated data of Ohnishi and Nitta (1959), and disregarding the small differences in numerical values of α, one obtains about 23 kcal/mole for the temperature region 301 - 328 K For a similar case, from the results of present work one obtains at least 30 kcal/mole for the temperature region 346 - 365 K. Despite the difference in PMMA and experimental conditions this illustrates the increase of activation energy with approach to the glass transition region.

Hoping that the above presented examples clearly indicated the applicability of the time-dependent rate constant (4.4) to describe adequately the kinetic patterns of radiation produced radicals in polymer matrices it is worth adding that variation with time of the form (4.4) also solves the problem of the so called step-wise decay much discussed for polymer systems several years ago, cf. e.g. Aulov and Sukhov (1973). As

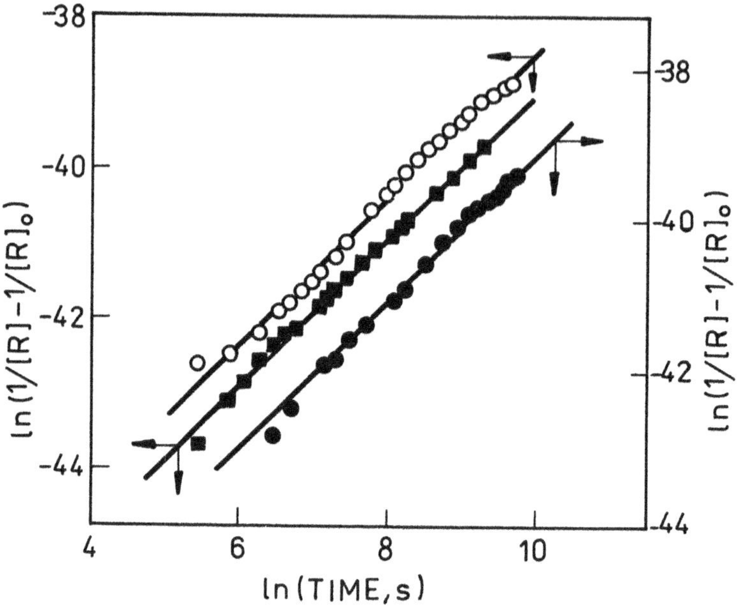

Fig. 4.10. Experimental data from Fig. 4.6 (●) and Fig. 4.9 (○,■) plotted according to relation (4.5)

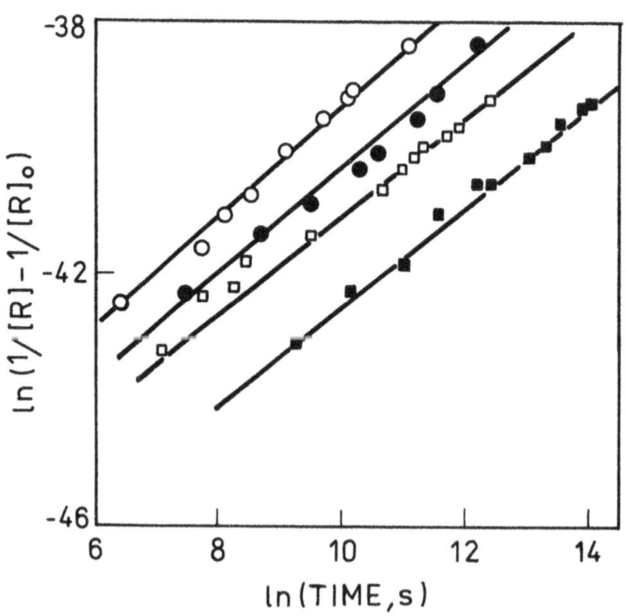

Fig. 4.11. Experimental data of Ohnishi and Nitta (1959) plotted according to relation (4.5); decay of propagating radicals at 301 (■), 311 (□), 318 (●), and 328 K (○)

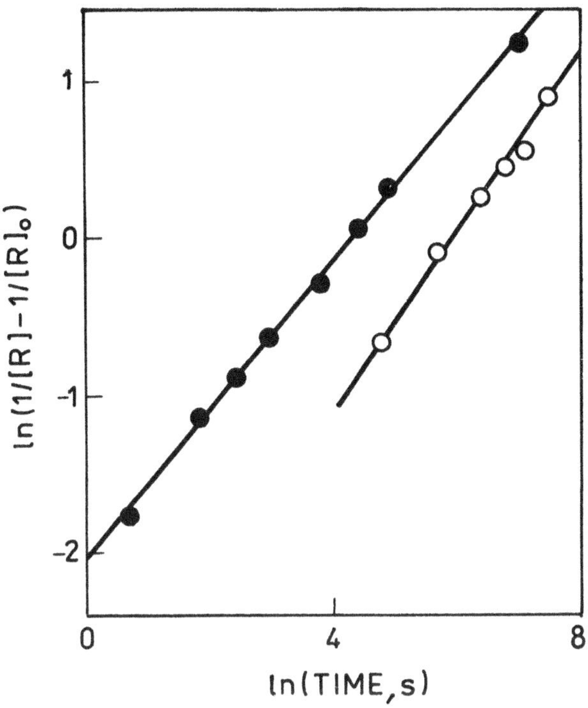

Fig. 4.12. Experimental data of Ormerod and Charlesby (1964) (●) and Pasal'skii et al. (1975) (○) plotted according to eqn. (4.5)

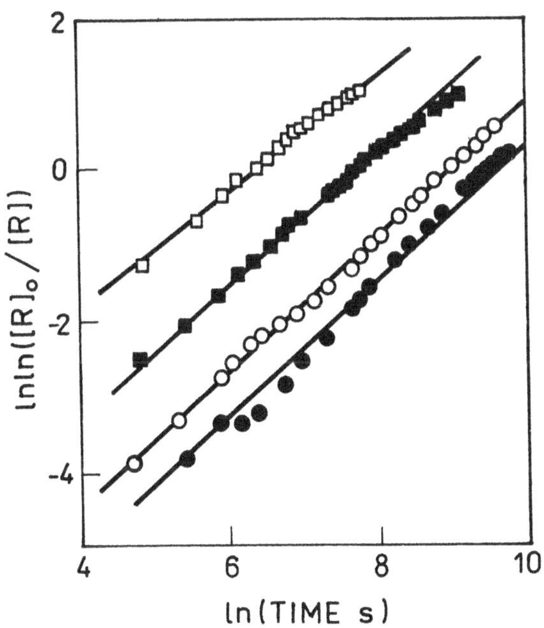

Fig. 4.13. Experimental data from Fig. 4.7 (□) and Fig. 4.8 (■,○,●) plotted according to eqn. (4.3)

Table 4.4. Rate parameters for propagating radical decay in γ-irradiated PMMA

PMMA	Irradiation[a,b]			Decay		B^c	α
$\bar{M}_v \cdot 10^{-5}$	Temp. K	Time h	Dose rate Mrad/h	Temp. K	Order		
Powder [d] 3.2	77[a]	-	-	283	II	$1.90 \cdot 10^{-21}$	0.57
Powder 6.5	77[a]	-	0.6	298	II	-	0.47
Block 6.5	77[b]	3.25	0.42	345	I	$1.41 \cdot 10^{-4}$	0.88
	77[b]	3.25	0.42	365	I	$2.38 \cdot 10^{-4}$	0.88
	77[b]	3.25	0.42	363	I	$11.1 \cdot 10^{-4}$	0.90
Powder 6-8	298[a]	-	0.71	301	II	$2.1 \cdot 10^{-23}$	0.79
	298[a]	-	0.71	313	II	$8.8 \cdot 10^{-23}$	0.77
	298[a]	-	0.71	318	II	$18.4 \cdot 10^{-23}$	0.84
	298[a]	-	0.71	328	II	$48.9 \cdot 10^{-23}$	0.87
Block 6.5	298[b]	0.50	1.20	346	II	$6.76 \cdot 10^{-22}$	0.96
	298[b]	0.50	1.20	353	II	$3.00 \cdot 10^{-21}$	1.00
	298[b]	0.50	1.20	365	I	$4.20 \cdot 10^{-3}$	0.75
Block 6.5	343[b]	2.0	0.49	353	II	$14.7 \cdot 10^{-22}$	1.00
	353[b]	2.0	0.49	353	II	$7.04 \cdot 10^{-22}$	0.96
	360	2.0	0.49	353	II	$5.44 \cdot 10^{-22}$	0.96

[a] Irradiation in vaccum, [b] in air, [c] first-order rate constant in $s^{-\alpha}$, second-order rate constants in g/spin $s^{-\alpha}$, [d] isotactic

it was shown in Chapter 3 there is nothing exceptional in this kind of decay pattern.

ITL OF X-IRRADIATED POLYSTYRENE

ITL of X-irradiated polystyrene was recently recognized (Hagekyriakou and Fleming 1982) as a first-order kinetic process with a distribution of electron trap activation energies. However, no attention was paid to the time-dependence of the isothermal decay of luminescence which should follow the kinetic pattern presented above for glassy matrices. Indeed, it is the case. The experimental results of Hagekyriakou and Fleming (1982) for polystyrene X-irradiated at equal dose rates for 30 s and 300 s are well fitted by eqn. (2.26), namely by

$$I(\tau,t) = 59.3 \cdot 10^3 t^{-0.66} \exp(-4.9 \cdot 10^{-2} t^{0.34}/0.34) \qquad (4.6)$$

for 30 s, cf. Fig. 4.14, and

$$I(\tau,t) = 65.9 \cdot 10^3 t^{-0.44} \exp(-1.03 \cdot 10^{-2} t^{0.56}/0.56) \qquad (4.7)$$

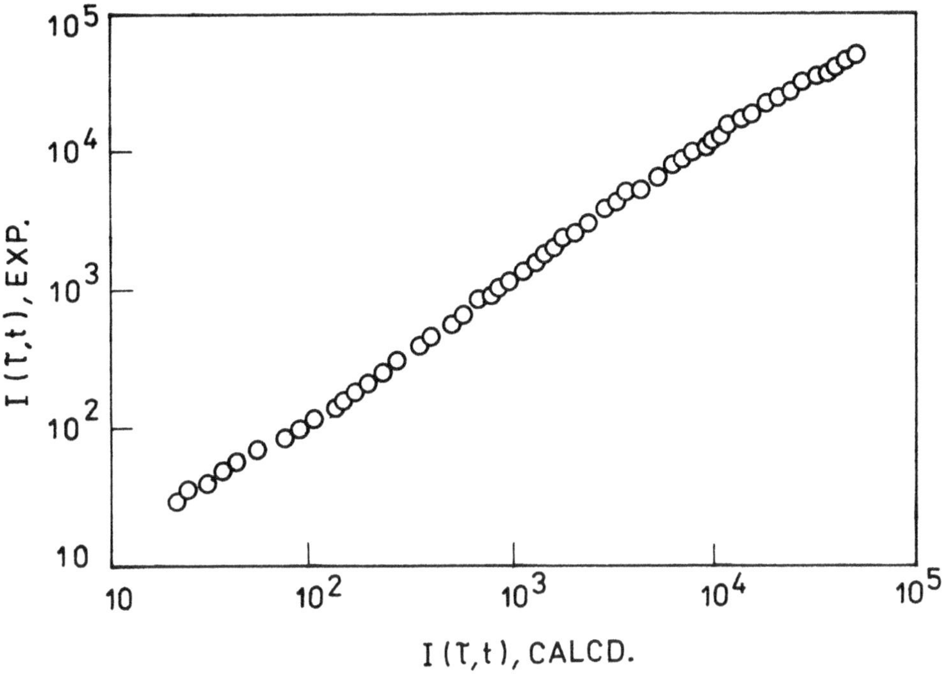

Fig. 4.14. Experimental data for ITL of polystyrene X-irradiated for 30 s at 130 K vs. numerical values calculated from eqn. (4.6)

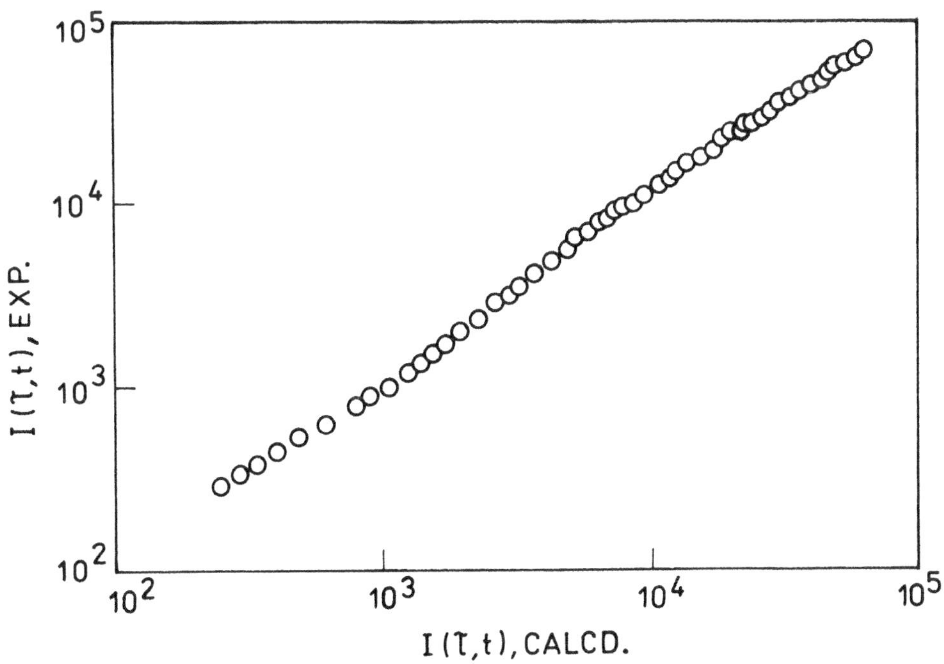

Fig. 4.15. Experimental data for ITL of polystyrene X-irradiated for 300 s at 130 K vs. numerical values calculated from eqn. (4.7)

for 300 s, cf. Fig. 4.15.

The distributions of electron lifetimes f(τ) and respective distributions of activation energies, g(E), calculated according to eqns. (1.6) and (1.8) for the rate parameters of eqns. (4.6) and (4.7) are depicted in Fig. 4.16. It is clearly seen that increase of irradiation time results in generation of electrons more firmly trapped in the matrix.

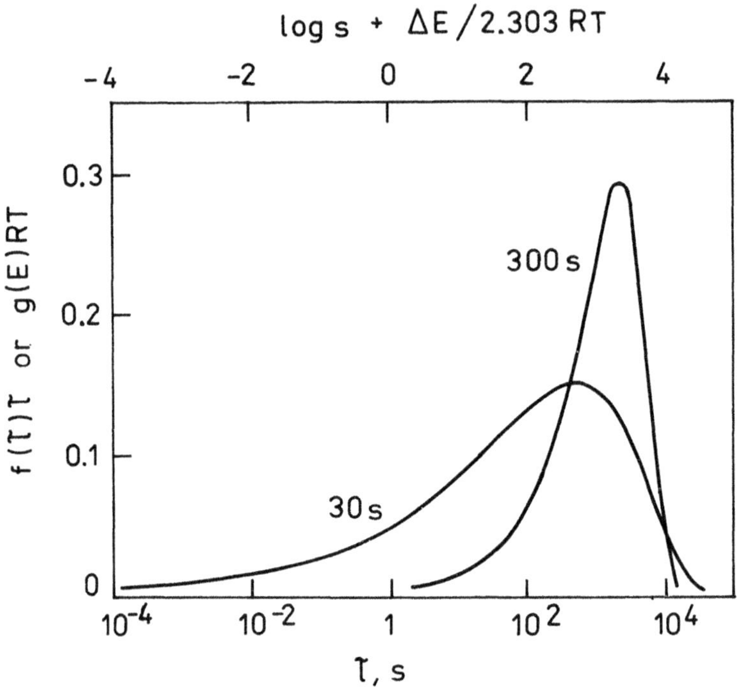

Fig. 4.16. Distributions of lifetimes, f(τ), and activation energies, g(E), of electrons involved in ITL of polystyrene X-irradiated for 30 and 300 s

REFERENCES

Auerbach I, Sanders LH (1969) Irradiated polyethylene. IV. Free radical decay. Polymer 10: 579-596

Aulov VA, Sukhov FF (1973) Isothermal decay of active products of polymer radiolysis. Step-wise decay of radicals (Russ). Khim Vys Energ 7: 412-417

Basheer R, Dole M (1983) Radiation chemistry of linear low-density polyethylene . II. Kinetics of alkyl and allyl free-radical decay reactions. J Polym Sci, Polym Phys Ed 21: 957-967

Butyagin PYu (1972) Decay of free radicals in polymer media. Pure Appl Chem 30: 57-76

Dole M, Hsu CS, Patel VM, Patel GN (1975) Kinetics of two simultaneous second-order reactions occuring in different zones. J Phys Chem 79: 2473-2479

Emanuel NM (1981) Chemical and biological kinetics (Russ). Usp Khim 50: 1721-1809

Hagekyriakou J, Fleming RJ (1982) Determination of the kinetic order of thermoluminescence in the presence of a distribution of electron trap activation energies. J Phys D, Appl Phys 15: 163-176

Hwang L-P, Cheng J-T (1981) Cage effects in recombination of allyl radical in irradiated polyethylene. J Polym Sci, Polym Phys Ed 19: 983-989

Ohnishi S-J, Nitta I (1959) Rate of formation and decay of free radicals in γ-irradiated polymethyl methacrylate by means of electron spin resonance absorption measurements. J Polym Sci 38: 451-458

Ormerod MG, Charlesby A (1964) Radiation chemistry of poly(methacrylic acid), poly(acrylic acid), and their esters - an electron spin resonance (ESR) study. Polymer 5: 67-88

Pasal'skii BK, Lavrentivich Yal, Vonsyatskii VA, Kabakchi AM (1975) Action of γ-radiation and α-particles on poly(methyl methacrylate) at -196° C (Russ). Khim Vys Energ 9: 483-486

Patel VM, Patel GN, Gvozdic N, Hsu CS, Dole M (1978) Radiation chemistry of polyethylene. XIV. Allyl radical decay kinetics in different types of polyethylene. J Polym Sci, Polym Phys Ed 16: 467-484

Plonka A (1983) Time-dependent reactivity of species trapped in polymer matrices: decay of allyl radicals in irradiated polyethylene. J Polym Sci, Polym Phys Ed 21: 1011-1016

Plonka A, Pietrucha K (1983) Kinetics of free-radical decay in γ-irradiated poly(methyl methacrylate). Radiat Phys Chem 21: 439-444

Szöcs F, Placek J, Kotlarik B (1977) Correlation of molecular motion and free-radical decay in irradiated PMMA. J Appl Polym Sci 21: 859-861

Waite TR (1958) General theory of bimolecular reaction rates in solids and liquids. J Chem Phys 28: 103-106

5. LIFE-TIME DISTRIBUTIONS OF PHOTOPRODUCED CATION-RADICALS IN MICELLES

A convenient model to study some features of photoionization processes which are pertinent for solar energy storage and relevant to biological electron-transfer processes is the aqueous solution of micelles with hydrophobic solute incorporated into them. The photoionization event occurs in the hydrocarbon-like region of the micellar core and the photoelectron has only to cross a short region of hydrocarbon phase in order to reach the micelle-water interface where it becomes hydrated (Wallace et al. 1973). This is schematically shown in Fig. 5.1 for a hydrophobic molecule incorporated into an anionic micelle sketched according to Menger proposition (Menger 1979; Menger and Dole 1984) to emphasize features such as rough surface, water filled pockets, looping, nonradial distribution of chains, and random distribution of therminal methyls.

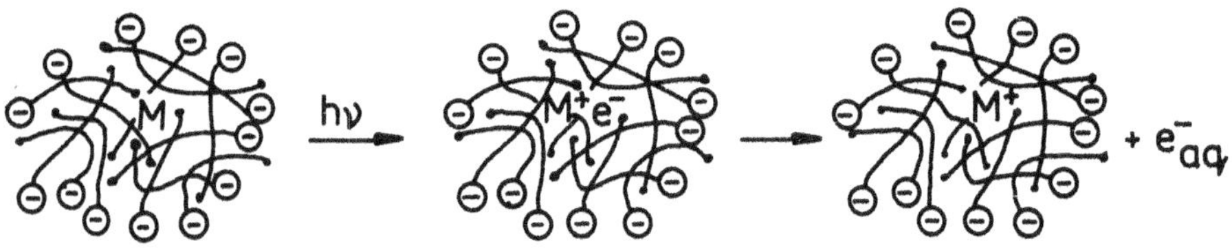

Fig. 5.1. Schematic representation of a photoionization process of a hydrophobic molecule M incorporated into an anionic micelle

In the above system the back reaction, i.e. recombination of photoelectron with parent cation is effectively suppressed by negatively charged interface and the photoionization is much more efficient than in homogeneous solutions (Kalyanasundaram 1978). This point is pertinent for solar energy utilization as the key aspect of any photoredox reaction for solar energy storage is to achieve net charge separation.

The fate of photoelectron as well as the fate of the parent cation are governed by the organization of the micellar system. Hence the biological interest of the micellar model: it forces the organization of reactants on the molecular level and mimics a simplified facet of biomembranes with an interface between aqueous phase and hydrophobic region. What we are interested in is the time-dependent reactivity of the parent cation and photoelectron.

TIME-DEPENDENT REACTIVITY OF SPECIES EMBEDDED IN MICELLES

Studies on the alternation of reaction order and reaction rates of species in micellar environments compared to homogeneous solution have shown time-dependent reactivity consisting of multiexponential patterns for first-order reactions (cf. Fendler 1982) and of reaction rate dependence on the storage time between the preparation of the micellar solution and the performance of an experiment (Grätzel and Thomas 1973). Let us illustrate this taking as an example the photoproduced N,N,N',N'-tetramethylbenzidine cation radical (TMB^+)

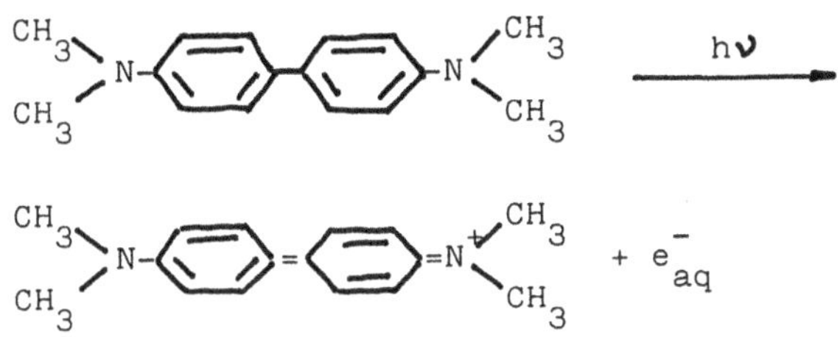

which in sodium dodecyl (SDS) micelles is easily followed by the ESR method, cf. Fig. 5.2 for the ESR spectrum (Narayana et al. 1981).

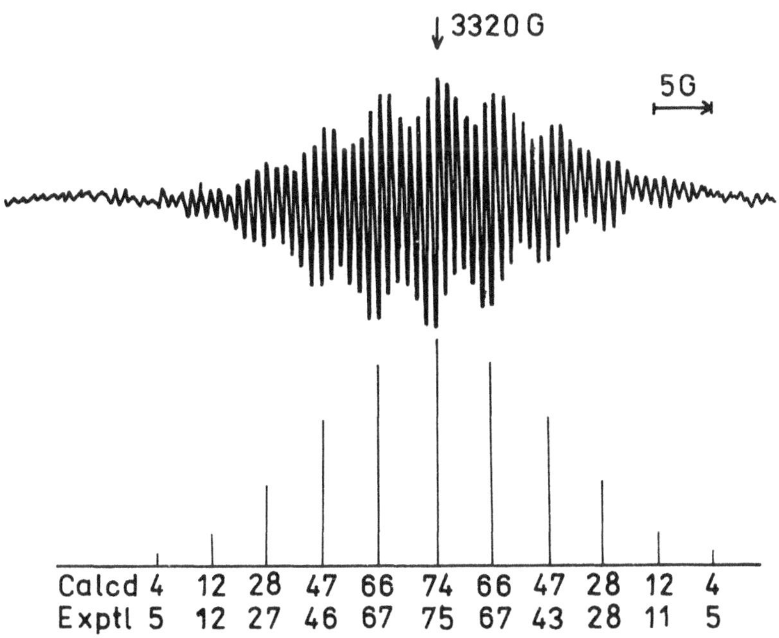

Fig. 5.2. ESR spectrum of TMB^+ in aqueous micellar solution of SDS

Plonka and Kevan (1984a) using a screen wire filters of transmittances down to 0.05 produced varying amounts of TMB^+ in freshly prepared 0.1 M solution of SDS by in situ irradiation during the same irradiation time of 100 s and followed the subsequent TMB^+ decay at the irradiation temperature of 305 K by recording the overmodulated ESR spectra. The results are presented in Fig. 5.3 in the coordinate system $\ln([TMB^+]/[TMB^+]_0)$ vs. time t where $[TMB^+]_0$ denotes the initial concentration of TMB^+ determined by extrapolation back to the end of the irradiation time from which the time t was counted. All experimental points fit the same multiexponential curve which indicates unambiguously a first-order decay with an apparent rate constant k(t) decreasing with time

$$-d[TMB^+]/dt = k(t)[TMB^+] \tag{5.1}$$

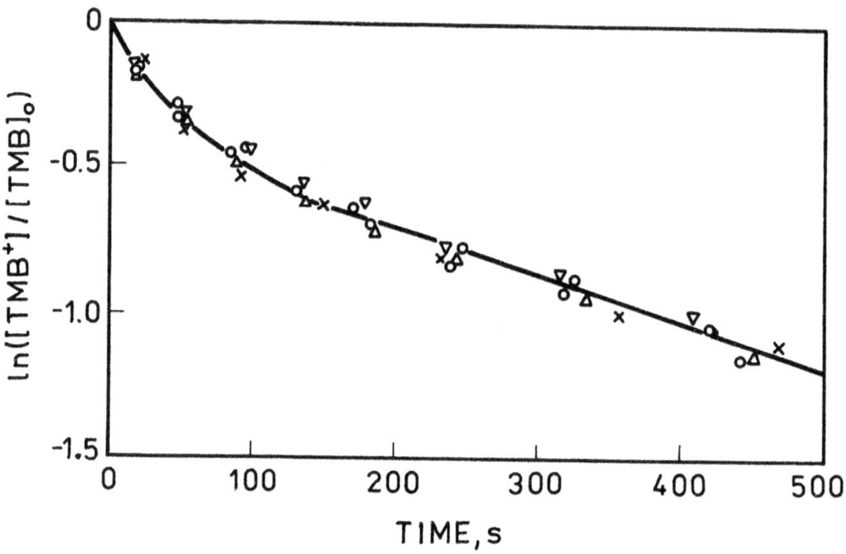

Fig. 5.3. Decay of photoproduced TMB^+ in freshly prepared 0.1 M SDS micellar solution

Assuming, as it was done for the systems discussed before, that the time dependence of k(t) can be expressed by

$$k(t) = Bt^{\alpha-1}, \qquad 0 < \alpha \leq 1 \tag{5.2}$$

one obtains by integrating eqn. (5.1)

$$[TMB^+]/[TMB^+]_0 = \exp(-Bt^{\alpha}/\alpha) \equiv \exp[-(t/\tau_0)^{\alpha}] \tag{5.3}$$

which is fitted perfectly by the experimental data , cf. Fig. 5.4, with α = 0.6.

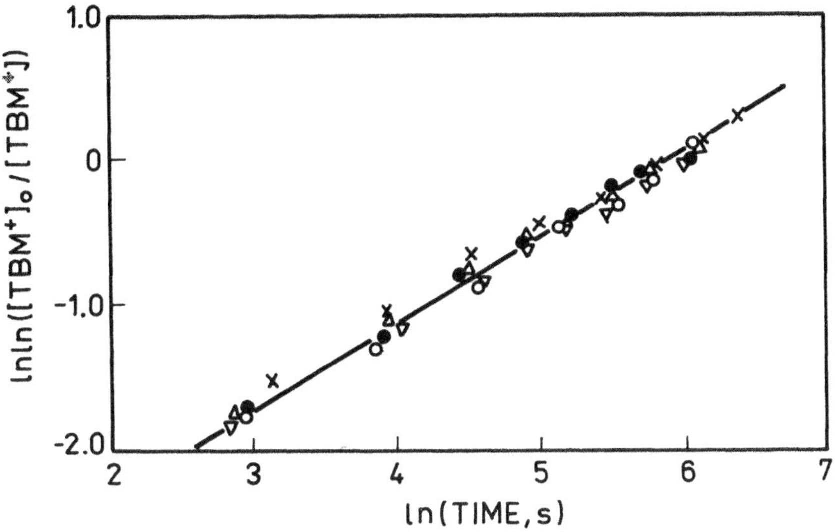

Fig. 5.4. Experimental data from Fig. 5.3 plotted according to eqn. (5.3)

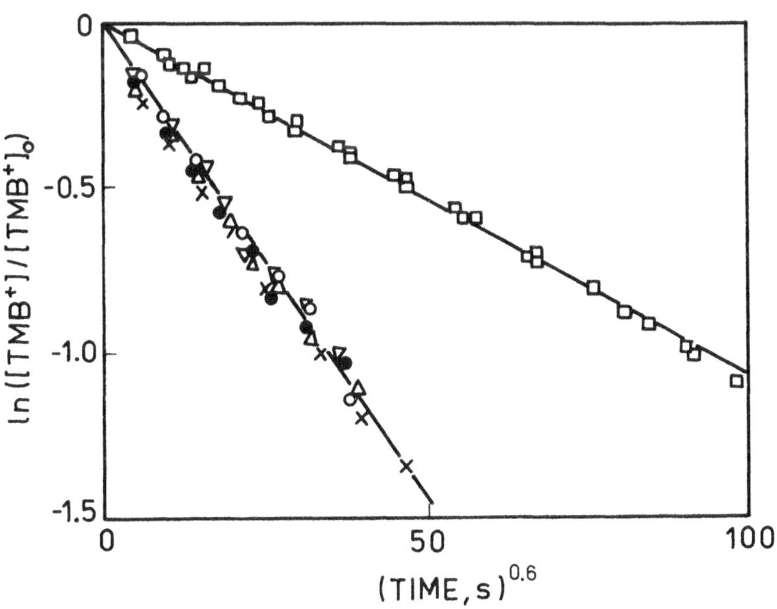

Fig. 5.5. Comparison of photoproduced TMB^+ decay in freshly prepared solutions, data from Fig. 5.3, and in stored for 1 week solutions of 0.1 M SDS (□)

Repeating the experiments under similar conditions for micellar solution stored for various times at room temperature before ultraviolet irradiation there was observed the decrease of the numerical values of B for virtually unchanged values of α. During storage for one week the numerical values of B decreased from 0.039 to 0.010 and no further changes were observed. This is illustrated in Fig. 5.5 in which the data are presented in the coordinate system $\ln([TMB^+]/[TMB^+]_o)$ vs. t^α for $\alpha = 0.6$.

The use of the first-order kinetic equation in this case can be rationalized as follows. Spectroscopic studies suggest that the probable mechanism of TMB^+ decay is disproportionation as only half of the photoionized TMB molecules are recovered after TMB^+ decay (Arce and Kevan 1985). In homogeneous solutions this disproportionation would be described by second-order equal concentration kinetics. In micellar solutions the rate determining step is not the disproportionation reaction itself but appears to be a cooperative process involving a correlation of the TMB^+ cation radicals disposed for reaction.

Turning to the microscopic origin of multiexponential decay we recall the interpretation in terms of a distribution of environments. However in the case of species embedded into micelles this picture seems to have more formal than physical significance since there is no effect of light intensity or irradiation time on the decay kinetics which might be expected to alter such a distribution. An alternative picture advanced recently is that nonexponential decay behaviour may arise from complex cooperative processes. We will return to this point in later sections. Here, the last point to be discussed is the effect of storage on the decay of TMB^+. One possible explanation (Grätzel and Thomas 1973) is that a completely equilibrated micelle may have a different alignment of the surfactant molecules than a freshly prepared one which could affect the separation of the charged head groups and increase the viscosity of the micellar interior.

SOLVENT ISOTOPE EFFECT

To illustrate the solvent isotope effect on the reactivity of species embedded in micelles, Fig. 5.6 shows the multiexponential decay patterns of photoproduced TMB^+ cation radicals in aged micelles of SDS in light and heavy water (Plonka and Kevan 1984b).

Data from Fig. 5.6 when plotted according to eqn. (5.3) in the coordinate system $\ln\ln([TMB^+]_o/[TMB^+])$ vs. $\ln t$, cf. Fig. 5.7, clearly display the significant difference between the numerical values of α for H_2O and D_2O solutions equal to about 0.6 and 0.4, respectively.

These numerical values of the constant α were found to be constant at 306 K despite the change from sodium dodecyl sulphate (SDS) to sodium decyl sulphate (SDeS), cf. Figs. 5.8 and 5.9.

The difference of the numerical values of α for H_2O and D_2O leads to a marked difference in lifetime distributions calculated from eqn. (1.6), cf. Fig. 5.10. The differences between the distributions of TMB^+ lifetimes in micelles of SDeS and SDS, either in H_2O or D_2O are small compared

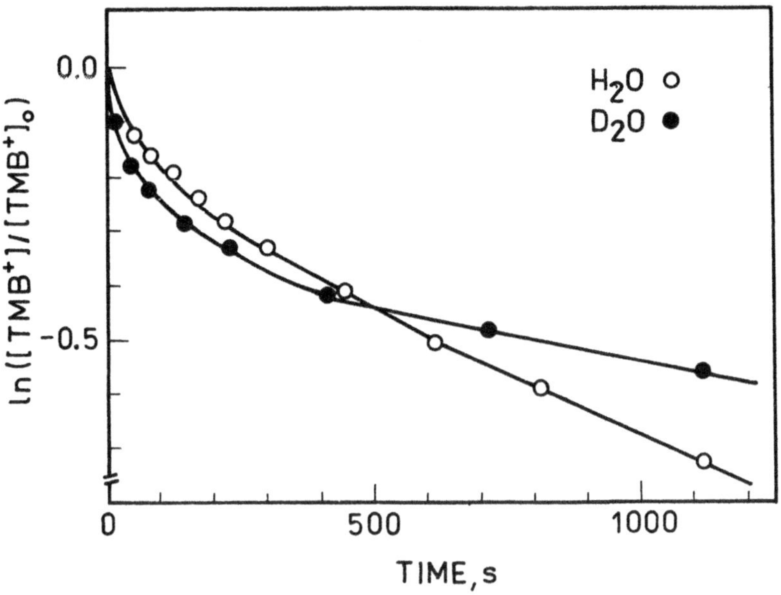

Fig. 5.6. Decay of photoproduced TMB^+ cation radicals at 306 K in aged solutions of SDS in H_2O and D_2O

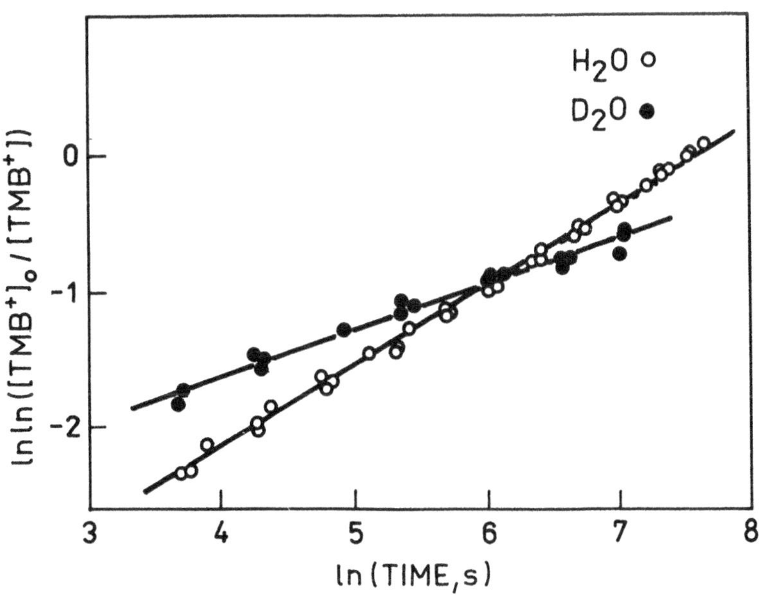

Fig. 5.7. Data from Fig. 5.6 and two other experiments performed under similar conditions plotted according to eqn. (5.3)

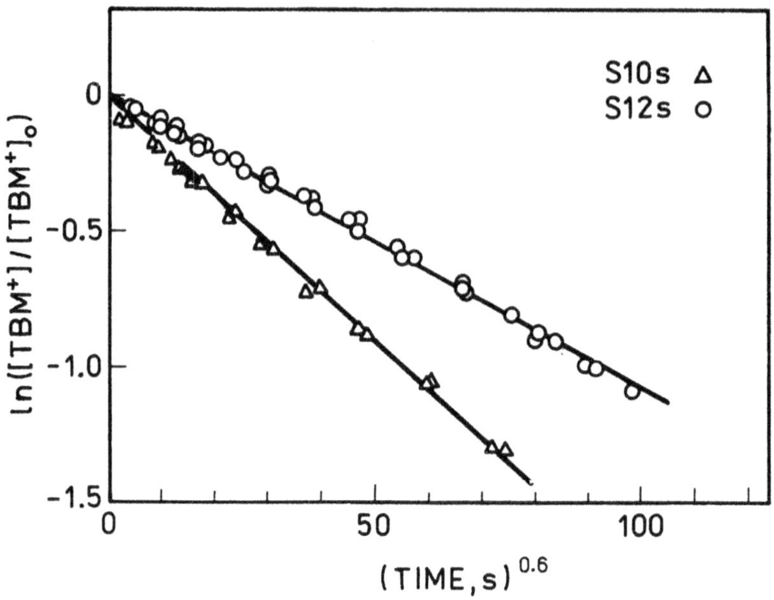

Fig. 5.8. Decay of photoproduced TMB^+ cation radicals in micelles of SDeS and SDS in H_2O

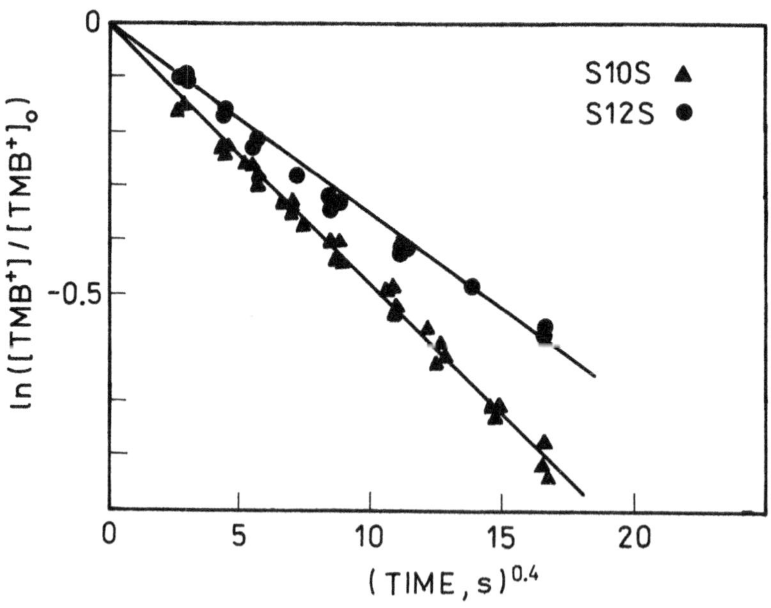

Fig. 5.9. Decay of photoproduced TMB^+ cation radicals in micelles of SDeS and SDS in D_2O

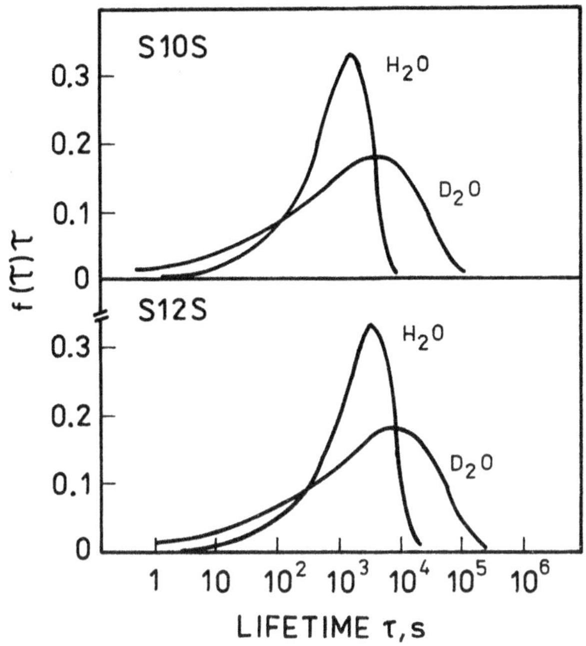

Fig. 5.10. Lifetime distributions of photoproduced TMB^+ cation radicals in micelles of SDeS and SDS in H_2O and D_2O

with that caused by the substitution of D_2O for H_2O. In a given medium the increase of alkyl chain length increases only the effective lifetime of TMB^+ cation radicals in micelles in agreement with conclusions on asymmetric localization of the solute within the micelles (Narayana et al. 1982). As the alkyl chain length increases the TMB cation radicals have decreased water interactions at the micellar interface and decay more slowly. Thus the maximum probable lifetime changes but the internal cooperative processes in the micelles are not changed and hence the lifetime distribution width of TMB cation radicals is not altered.

Substitution of D_2O for H_2O not only increases the effective lifetime of TMB^+ but also causes a significant broadening of TMB^+ lifetime distribution. This lifetime broadening effect, similar for both kinds of micelles, can arise from tighter molecular packing inside the micelle in D_2O than in H_2O due to stronger hydrophobic interactions in D_2O which has been demonstrated for a number of lipid aqueous dispersions (Chen 1982).

Tighter molecular packing makes the micelle dynamics more cooperative and leads to a wider distribution of lifetimes (Skinner 1983). This tighter molecular packing in D_2O also explains why the effect of alkyl chain length on the effective lifetime of TMB cation radicals is slightly greater in H_2O than in D_2O. Additional support for cooperative processes in micelle dynamics leading to noneponential decay of embedded species comes from the fact that their effective lifetimes are many orders of magnitude greater than the time scale of events for solute exists from micelles or for the exchange of surfactant molecules between the micelle and the bulk solution (Thomas et al. 1978; Turro et al. 1980).

EFFECT OF MICELLAR INTERFACE MODIFICATIONS ON REACTIVITY OF EMBEDDED SPECIES

The self-association of amphiphiles in aqueous solutions into micellar aggregates is well recognized (Tanford 1980) as a cooperative process requiring simultaneous participation by many amphiphilic molecules or ions. This process is driven by the transfer of hydrocarbon chains from a polar solvent to the less polar interior of a micelle, the hydrophobic effect, and opposed by the repulsion among the polar headgroups when they are brought into close contact at the surface of the micelle. Having explored the effect of hydrophobic interaction on reactivity of species embedded in micelles we turn now to modifications of micellar interface. Making use of the finding that in the presence of anionic electron scavengers the yield of photoproduced TMB^+ radicals in cationic micelles of dodecyltrimethylammonium chloride (DTAC) increases to the level at which its ESR spectrum can be detected at room temperature, Plonka and Kevan (1985a) were able to compare TMB^+ decay in 0.1 M micellar solutions of DTAC (in the presence of $5 \cdot 10^{-3}$ M of $NaNO_3$), SDS, and tetramethylammonium dodecyl sulfate (TMADS), cf. Fig. 5.11.

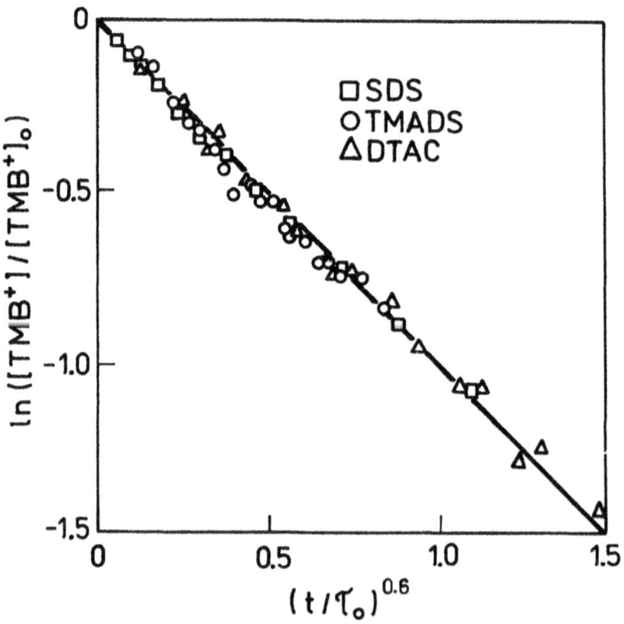

Fig. 5.11. Decay at 306 K of photoproduced TMB^+ in 0.1 M micellar solutions of SDS, TMADS and DTAC

It is seen that the experimental data fit well eqn. (5.3) with a common value of α = 0.6, τ_0 being equal to $2.0 \cdot 10^3$, $1.4 \cdot 10^3$, and $0.30 \cdot 10^3$ s for micellar solutions of SDS, TMADS, and DTAC, respectively. From this it follows that TMB^+ lifetime distributions in these micelles have the same width, as indicated by the common value of α , and the only effect of the variation of the counterion or the headgroups is to change the position on the lifetime scale.

Qualitatively similar effects were observed upon addition of additives such as salt or alcohol which are used to modify the micellar surface potential. Neither addition of sodium chloride or 1-butanol change the width of the TMB^+ lifetime distribution in SDS micelles.

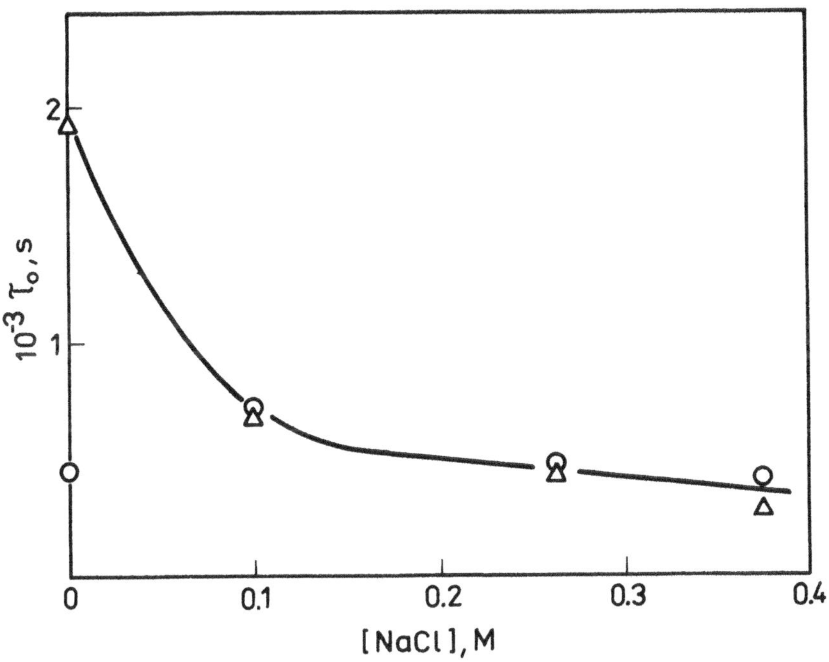

Fig. 5.12. Effect of sodium chloride on the effective lifetime at 306 K of photoproduced TMB^+ in freshly prepared (○) and stored for one week (△) 0.1 M micellar solutions of SDS

Both additives only decrease the effective lifetime. In the case of sodium chloride, cf. Fig. 5.12, the effect of salt on the aging of the micellar solution of SDS is illustrated. It appears that the presence of salt promotes the aging process since there is no differences in lifetime for freshly prepared and stored SDS solutions containing NaCl. The effect of 1-butanol on the aging of micellar solutions of SDS was much less evident. It is interesting to note, however, that there is no difference in lifetime for micellar solutions stored before addition of 1-butanol and solutions stored with added alcohol, cf. Fig. 5.13.

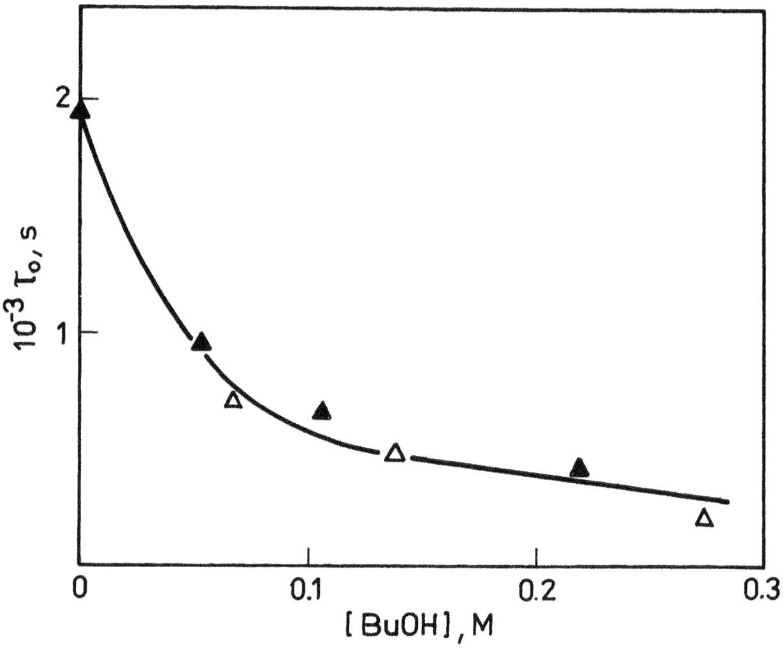

Fig. 5.13. Effect of 1-butanol on the effective lifetime at 306 K of photoproduced TMB^+ in 0.1 M micellar solutions of SDS stored for 1 week before addition of alcohol (▲) and with alcohol added (△)

From recent ESR and ESEM studies of TMB photoionization in anionic micelles it was found (Szajdzinska-Pietek et al. 1984) that substitution of tetramethylammonium counterion for sodium counterion significantly influences the structure of SDS micellar solutions which in turn affects both the photoionization efficiency of a micellar solubilizate and the stability of the photoproduced cations. The explanation given was that sodium counterions are located in the outer part of the headgroup region so that a relatively compact headgroup structure is maintained in SDS micelles. The more hydrophobic tetramethylammonium cations are located in the inner part of the headgroup region acting as spacers between the headgroups and create a less compact headgroup structure with a higher local concentration of water. The stronger TMB^+ radical-water interaction in TMADS micelles compared to SDS micelles explains fairly well both the higher photoionization efficiency of TMB and TMADS, since deeper water penetration assists in solvation of photoejected electrons which promotes charge separation, and the lower stability of photoproduced TMB^+ radicals due to less efficient embedding. The interesting point is that the width of TMB^+ lifetime distributions is not changed by the change in counterion. The lifetime distribution width is also the same in anionic and cationic micelles even through the effective lifetime is much less in cationic micelles.

There seems to be a simple rationalization of the width invariance of TMB^+ radical lifetime distribution for micelles prepared in H_2O. Variation of headgroup or counterion have no effect on the hydrophobic properties of the alkyl chain on which the constrained concerted motion of hydrocarbon chains depends. This assumption is implied in Stigter's (1974a,b) treatment of micelle formation by ionic surfactancts and by ^{13}C spin-lattice relaxation time measurements for micelles of SDS (Roberts and Chachaty 1973) and dodecylammonium chloride (Kalyanasundaram et al. 1975). Hence the cooperativity of micellar dynamics is not changed and the width of the TMB^+ radical lifetime distribution is not affected. Variations of headgroup or counterion do affect, however, the TMB^+ radical-water interaction which appears to be one of the main factors determining the effective lifetime.

The above conclusions are supported by the results obtained for SDS micellar solutions with salt or alcohol added to modify the micellar surface potential (Grand et al. 1983). The presence of an electrolyte in an aqueous SDS solution neutralizes the surface charge and decreases the electrostatic repulsion between the headgroups. For high salt concentration it is expected that SDS micelles would behave like nonionic ones. Upon NaCl addition (Stigter 1974a; Hayashi and Ikeda 1980; Lianos and Zana 1980; Almgren and Löfroth 1981; Croonen et al. 1983) the critical micelle concentration decreases and the aggregation number increases yet the TMB^+ radical lifetime distribution retains the same width and the effective lifetime decrease. The decrease of TMB^+ effective lifetime correlates fairly well with the increased TMB^+-water interaction as deduced from ESEM experiments. It was found (Maldonado et al. 1984) that TMB^+-water interactions increase with salt addition to a maximum at about 0.2 M and then decrease somewhat. For SDS micelles with added NaCl measurements have been made of the ^{13}C nuclear magnetic resonance longitudinal relaxation time of the surfactant molecule and it was found (Roberts and Chachaty 1973) that up to about 0.3 M NaCl the molecular motions inside the micelle did not change. This seems compatible with an unchanged distribution width of TMB^+ lifetimes.

Alcohol molecules intercalate between surfactant ions (Larsen and Tepley 1974; Zana et al. 1981; Jain and Singh 1981; Almgren and Swarup 1983) and increase the average distance between ionic headgroups which decreases the surface charge density and consequently increases the degree of micellar dissociation. The critical micelle concentration and the aggregation number decrease yet the width of the TMB^+ radical lifetime distribution remains unchanged and the effective lifetime decreases. In micellar solutions of SDS, TMB^+-water interactions increase upon addition of n-butanol up to 0.1 M and remains constant with a further increase of alcohol content (Szajdzinska-Pietek et al. 1985). Once more the effective lifetime of TMB^+ radicals correlates fairly well with TMB^+-water interactions as deduced from ESEM experiments. Thus in a given micellar system cation radical - water interactions appear to be the main factor determining the effective lifetime of the embedded TMB^+ radicals.

DISTRIBUTION OF ACTIVATION ENERGY FOR DECAY OF EMBEDDED SPECIES

The decays of photoproduced TMB^+ radicals in aged 0.1 M micellar solutions of SDS, with and without 0.1 M NaCl, were followed at irradiation temperatures in the range 287 - 328 K. Micellar solutions with and without NaCl added were compared because it was seen above that in the presence of electrolyte there is an acceleration of micelle aging as well as a change in the photoproduced TMB^+ location within the micelles (Maldonado et al. 1984). The results are presented in Figs. 5.14 and 5.15 in the coordinate system $\ln \ln([TMB^+]_o/[TMB^+])$ vs. ln t in which eqn. (5.3) yields a straight line.

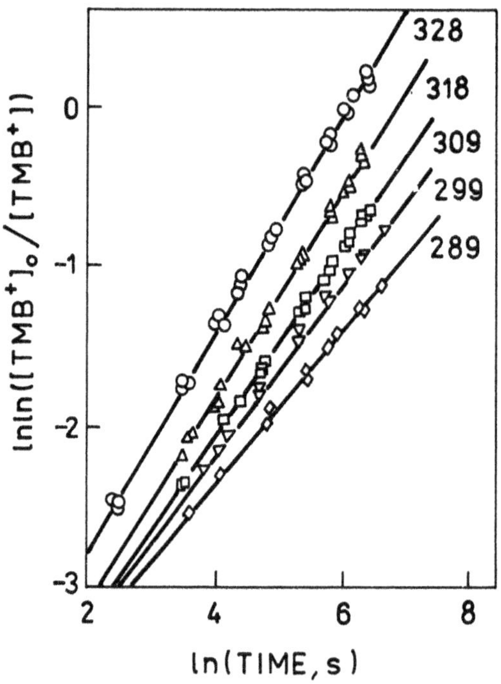

Fig. 5.14. Decay of photoproduced TMB^+ radicals in 0.1 M SDS micellar solutions plotted according to eqn. (5.3)

For TMB^+ in SDS micellar solutions without added NaCl the constant α changes in the range from 0.5 to 0.7 while in solutions with added NaCl it remains about constant at 0.6. Because of the variation in α it is necessary to use the mean lifetime $\bar{\tau}$ for the $f(\tau)$ distribution, cf. Appendix A,

$$\bar{\tau} = (\tau_o/\alpha)\,\Gamma(1/\alpha) \qquad (5.4)$$

in Arrhenius relation

$$\bar{\tau} = \exp(\bar{E}/RT) \qquad (5.5)$$

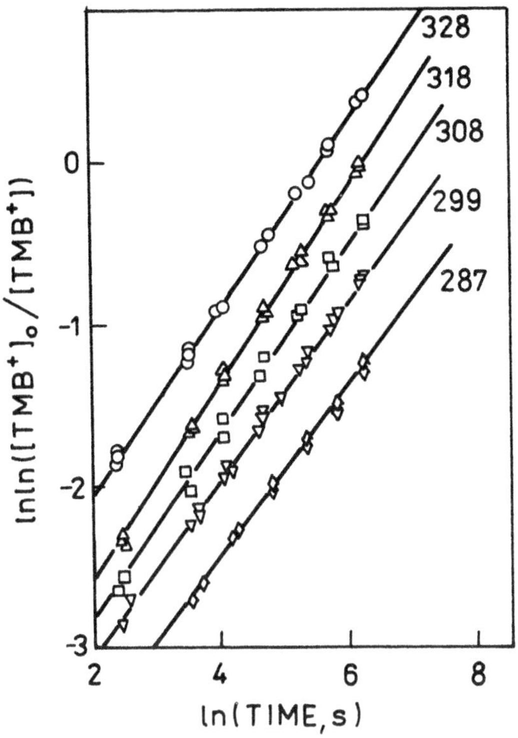

Fig. 5.15. Decay of photoproduced TMB^+ radicals in 0.1 M SDS micellar solutions with 0.1 M NaCl added, plotted according to eqn. (5.3)

to characterize the temperature dependence of the TMB^+ decay. This gives the results plotted in Fig. 5.16.

It is evident that the experimental results fit quite well to the Arrhenius relation with $\bar{E}$ equal to about 66 kJ $mole^{-1}$ for SDS micellar solutions without added NaCl and to 59 kJ $mole^{-1}$ for SDS micellar solutions with added NaCl. This difference of $\bar{E}$ is within the experimental errors of about 10% and is consistent with the conclusions that modification of the micellar interface by addition of sodium chloride has little effect on the decay of embedded photoproduced cations. The electrolyte neutralizes the micellar surface charge and decreases the electrostatic repulsion between the headgroups but has little influence on the molecular motion inside the micelles, as it was seen above.

It is worth noting that the decrease of the constant α for SDS micellar solutions without added NaCl from about 0.7 at 328 K to about 0.5 at 289 K parallels the change from about 0.6 to about 0.4 observed at 306 K upon substitution of D_2O for H_2O in micellar solutions of SDS. The higher viscosity, heat capacity and temperature of maximum density of D_2O relative to H_2O suggests that D_2O is substantially more structured than H_2O

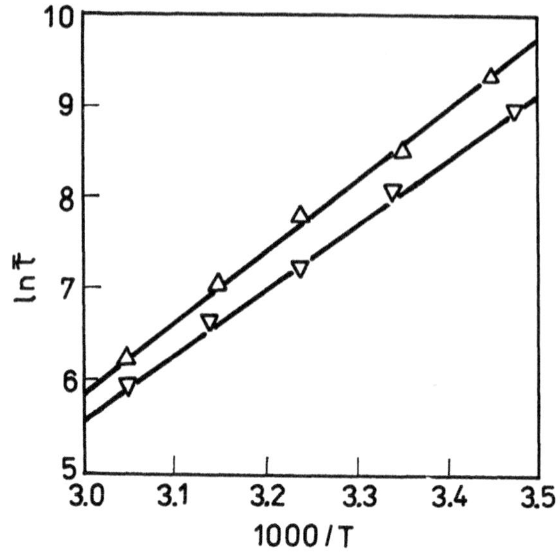

Fig. 5.16. Arrhenius plots for the temperature dependence of the mean lifetime of TMB^+ radicals in SDS micellar solutions with (▽) and without 0.1 M NaCl added (△)

at room temperatures and that temperature properties of D_2O can be compared to those of H_2O at a lower temperature, cf. Mukerjee et al. (1966).

The lifetime distributions for α equal to 0.5, 0.6 and 0.7 are presented in Fig. 5.17 in the coordinate system $f(\tau)\tau$ vs. $\ln(\tau/\bar{\tau})$. In considering the microscopic origin of the TMB^+ lifetime distribution it was pointed out (Plonka and Kevan 1985b) that movement of one alkyl tail of a surfactant molecule to allow for reaction of TMB^+ in SDS micelles depends on the configuration and dynamics of other hydrocarbon chains in the core of the micelle. Recent theoretical analysis (Blomberg and Edholm 1984) of the constrained concerted motions in hydrocarbon chains has revealed that because of its multidimensional character it is not sufficient to regard the motion across a saddle point in the multidimensional space but that a more extensive density of energy barriers must be considered. This leads to a distribution of activation energies, $g(E)$, which may be assumed to be the origin of $f(\tau)$. These activation energy distributions coincide with the corresponding TMB^+ lifetime distributions shown in Fig. 5.17, when plotted as $g(E)RT$ vs. $(E-\bar{E})/RT$ because

$$g(E) = f(\tau)(d\tau/dE) = f(\tau)\tau/RT \qquad (5.6)$$

and

$$\ln(\tau/\bar{\tau}) = (E - \bar{E})/RT \qquad (5.7)$$

if one assumes that the Arrhenius relation holds for any τ. These acti-

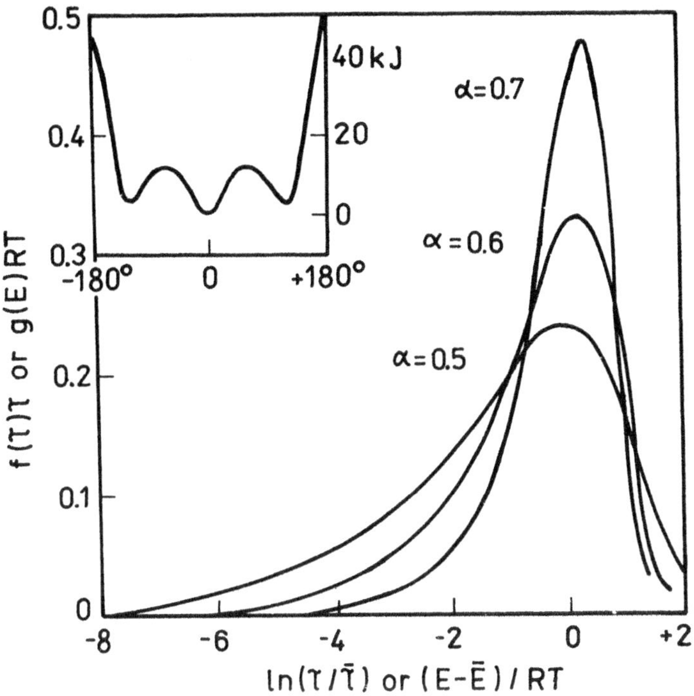

Fig. 5.17. The distributions of lifetime, $f(\tau)\tau$ and activation energies, g(E) for TMB^+ decay in SDS micelles. The insert shows a typical theoretical potential for bond rotation in a hydrocarbon chain, cf. Skolnick and Helfand (1980)

vation energy distributions g(E) depend on temperature. At 305 - 306 K, the temperature at which most of the experiments were performed, for α equal to about 0.6, g(E) extends for about 10 RT i.e. 25 kJ $mole^{-1}$ from about 45 to 70 kJ $mole^{-1}$ with a maximum at about 66 kJ $mole^{-1}$. The distribution is broader for lower numerical values of α , for α = = 0.5 by approximately 2 RT, and narrower for higher numerical values of α , for α = 0.7 by approximately 2 RT. These variations are within the estimated 10% experimental error of 66 kJ $mole^{-1}$ and the shifts of the distribution maxima are experimentally negligible. The insert in Fig. 5.17 shows a typical theoretical potential for bond rotation in an isolated carbon chain. This potential has maxima at 60° of 12.36 kJ $mole^{-1}$ and ±180° of 44.83 kJ $mole^{-1}$. For a constrained hydrocarbon chain there will be a distribution of potential barriers with an estimated maximum at about 1.3 times the value for isolated hydrocarbon chains (Blomberg and Edholm 1984). This is in good agreement with the experimentally derived activation energy distributions g(E) which show a maximum of 66 kJ $mole^{-1}$ which is about 1.5 times the theoretical potential maximum at 180° of 45 kJ $mole^{-1}$ for single bond rotation in isolated hydrocarbon chains. Thus, one may conclude that the experimentally observed activation energy distribution for TMB^+ decay in micelles is related to conformational changes involving rotation around internal bonds of hydrocarbon chains in micelles and accounts for the origin of the observed TMB^+ lifetime distributions. This interpretation of the activation energy distribution is consistent with cooperative dynamics which seem to be general in constrained environments. Recent theoretical work has shown that strong dynamic cooperativity can lead to nonexponential relaxation (Skinner 1983) which describes the decay kinetics observed for TMB^+ in micelles. Dynamic cooperativity can be described microscopically as dynamics controlled by the cooperative motion of environmental molecules rather than by the motion of an isolated reacting molecular species. Additional experimental support for dynamic cooperativity in micelles comes from the discussed effects of surfactant alkyl chain length and preparation of micelles in H_2O vs. D_2O on TMB^+ decay. As the alkyl chain length increases the cooperative processes in the micellar interior are not changed and hence the lifetime distribution width for TMB^+ decay is not changed. However, substitution of D_2O for H_2O causes significant broadening of the TMB^+ lifetime distribution due to tighter molecular packing of the micelle interior.

Nonexponential decay kinetics have often been more or less adequately approximated by a finite number, typically small, of exponential decays. However the physical interpretation of a small number of different but unspecified decay processes is often unclear. In constrained environments the distribution of decay processes provides a better physical description. This distribution is a reflection of microscopic dynamic cooperativity.

KINETIC BEHAVIOUR OF PHOTOEJECTED ELECTRONS

Solvation kinetics of electron photoejected from hydrophobic molecule incorporated in anionic micelle, cf. Fig. 5.1, was recently studied by Gauduel et al. (1984) by femtosecond time-resolved absorption spectroscopy. The upper limit for solvation time of electron in aqueous solutions was estimated to be about 100 fs and it was demonstrated that the molecular organization in SDS micelles induces a slight delay in solvation of electron photoejected from phenothiazine. This is seen in Fig. 5.18 reproducing the rise of the transient absorbance at 750 nm which is due to the hydrated electron.

What is much more interesting, the transient absorbance at 750 nm was shown to decay in several hundred picoseconds, cf. Fig. 5.19, according to

$$A_{750}/A^{o}_{750} = \exp\left[-(t/\tau_{o})^{\alpha}\right] \qquad (5.8)$$

i.e. according to the first-order kinetic equation with time-dependent rate constant (5.2) for $\alpha = 0.5$. Interpreting this result Gauduel et al. (1984) suggested that electrons disappear in reactions with impurities, like oxygen molecules, present in small concentrations in the aqueous phase. Taking the rhs of eqn. (5.8) as the random walk form they assume that electrons do diffuse out through the boundary layer before the reverse electron transfer to the inner oxidized cation (PTH^+) occurs. However, remembering that at about the same temperature the

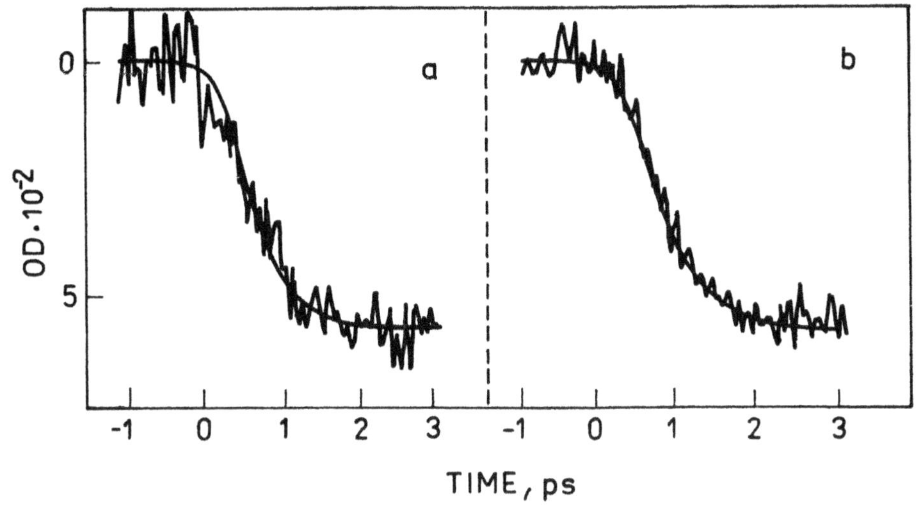

Fig. 5.18. Transient absorbance at 750 nm after subpicosecond laser photolysis of phenothiazine, PTH, in SDS micelles at about 293 K: a) $1.0\cdot 10^{-}$ M PTH in 0.025 M SDS, b) $1.5\cdot 10^{-4}$ M PTH in 0.1 M SDS. Solid lines represent computed best fits giving hydrated electron appearance time constant of 350 and 450 fs, respectively

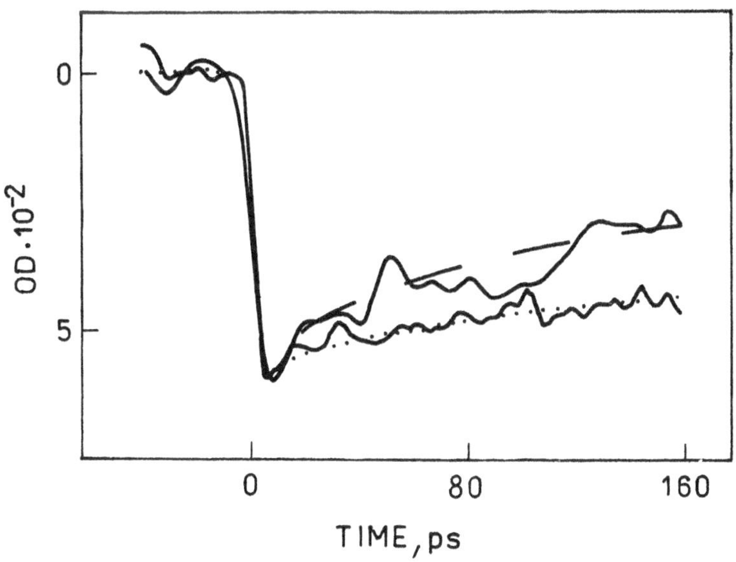

Fig. 5.19. Transient absorbance at 750 nm showing the decay at about 293 K of hydrated electron in SDS micellar solutions. The decays are fitted by eqn. (5.8) with τ_o = 1100 ps for $1.5\cdot 10^{-4}$ M in 0.1 M SDS (••••••) and τ_o = 250 ps for $1.0\cdot 10^{-4}$ M PTH in 0.025 M SDS (— —)

photoproduced TMB^+ radicals in SDS micelles do decay according to similar equation with the same numerical value of α , cf. Fig. 5.14, one can suggest that electron solvation proceeds somewhere in the water filled pockets of micelles, cf. Fig. 5.1, and their subsequent release into the aqueous phase is due to the same mechanism as in the case of embedded cation-radicals. The time scale are different but reactivity distributions are the same.

REFERENCES

Almgren M, Löfroth J-E (1981) Determination of micelle aggregation numbers and micelle fluidities from time-resolved fluorescence quenching studies. J Colloid Interface Sci 81: 486-499

Almgren M, Swarup S (1983) Size of sodium dodecyl sulfate micelles in the presence of additives. I. Alcohols and other polar compounds. J Colloid Interface Sci 91: 256-266

Arce R, Kevan L (1985) Photochemical behaviour of N,N,N',N'-tetramethylbenzidine and its protonated forms in sodium dodecyl sulfate anionic micelles under 337 nm laser irradiation . J Chem Soc, Faraday Trans I 81: 1669-1676

Blomberg C, Edholm O (1984) Analysis of constrained concerted motions in hydrocarbon chains with applications to kinks in lipid bilayers. Mol Phys 51: 485-498

Chen Ch-H (1982) Interactions of lipid vesicles with solvent in heavy and light water . J Phys Chem 86: 3559-3562

Croonen Y, Geladé E, van der Zegel M, van der Auweraer M, Vandendriessche H, De Schryve FC, Almgren M (1983) Influence of salt, detergent concentration, and temperature on the fluorescence quenching of 1-methylpyrene in sodium dodecyl sulfate with m-dicyanobenzene. J Phys Chem 87: 1426-1431

Fendler JH (1982) Membrane mimetic chemistry. Wiley Interscience, New York

Gauduel Y, Migus A, Martin JL, Antonetti A (1984) Femtosecond and picosecond time-resolved electron solvation in aqueous and reversed micelles. Chem Phys Lett 108: 319-322

Grand D, Hautecloque S, Bernas A, Petit A (1983) Effect of interfacial electric charge on photoionization yields in micellar system. J Phys Chem 87: 5236-5240

Grätzel M, Thomas JK (1973) On the dynamics of pyrene fluorescence quenching in aqueous ionic micellar systems. Factors affecting the permeability of micelles. J Am Chem Soc 95: 6885-6889

Hayashi S, Ikeda S (1980) Micelle size and shape of sodium dodecyl sulfate in concentrated NaCl solutions. J Phys Chem 84: 744-751

Jain AK, Singh RPB (1981) Effect of alcohols on counterion association in aqueous solutions of sodium dodecyl sulfate. J Colloid Interface Sci 81: 536-539

Kalyanasundaram K (1978) Photophysics of molecules in micelle-forming surfactant solutions. Chem Soc Rev 7: 453-472

Kalyanasundaram K, Grätzel M, Thomas JK (1975) Electrolyte-induced phase transitions in micellar systems. A proton and carbon-13 nuclear magnetic resonance relaxation and photochemical study. J Am Chem Soc 97: 3915-3922

Larsen JW, Tepley LB (1974) Effect of aqueous alcoholic solvents on counterion binding to CTAB (cetyltrimethylammonium bromide) micelles. J Colloid Interface Sci 49: 113-118
Lianos P, Zana R (1980) Use of pyrene excimer formation to study the effect of NaCl on the structure of sodium dodecyl sulfate micelles. J Phys Chem 84: 3339-3340
Maldonado R, Kevan L, Szajdzinska-Pietek E, Jones RRM (1984) Effect of added sodium chloride on the molecular environment and photoionization of N,N,N',N'-tetramethylbenzidine in micellar solutions as studied by electron spin echo and electron spin resonance spectroscopy. J Chem Phys 81: 3958-3961
Menger FM (1979) On the structure of micelles. Acc Chem Res 12: 111-117
Menger FM, Doll DW (1984) On the structure of micelles . J Am Chem Soc 106: 1109-1113
Mukerjee P, Kapauan P, Meyer HG (1966) Micelle formation and hydrophobic bonding in deuterium oxide. J Phys Chem 70: 783-786
Narayana PA, Li ASW, Kevan L (1981) Electron spin resonance and electron spin echo studies of photoproduced tetramethylbenzidine cation radical in frozen aqueous micellar solutions: cation surroundings and retention of micellar structure in frozen solutions. J Am Chem Soc 103: 3603-3604
Narayana PA, Li ASW, Kevan L (1982) Electron spin resonance and electron spin echo studies of the photoionization of tetramethylbenzidine in frozen aqueous anionic, cationic, and nonionic micellar solutions. Effect of micelle type and anionic micelle size. J Am Chem Soc 104: 6502-6505
Plonka A, Kevan L (1984a) Time dependent reactivity of species embedded in micelles: lifetime distribution of photoproduced N,N,N',N'-tetramethylbenzidine cation radicals in sodium dodecyl sulfate micelles. J Chem Phys 80: 5023-5026
Plonka A, Kevan L (1984b) Solvent isotope effect on reactivity of photoproduced cations embedded in micelles. J phys Chem 88: 6348-6350
Plonka A, Kevan L (1985a) Effect of micellar interface modifications on the reactivity of embedded photoproduced cations. J Phys Chem 89: 2087-2089
Plonka A, Kevan L (1985b) Temperature dependence of the nonexponential decay of photoproduced N,N,N',N'-tetramethylbenzidine cation radicals in sodium dodecyl sulfate micelles: determination of activation energy distributions and relation to dynamic cooperativity. J Chem Phys 82: 4322-4325
Roberts RT, Chachaty C (1973) ^{13}C relaxation measurements of molecular motion in micellar solutions. Chem Phys Lett 22: 348-351
Skinner JL (1983) Kinetic Ising model for polymer dynamics: Applications to dielectric relaxation and dynamic depolarized light scattering. J Chem Phys 79: 1955-1964
Skolnick J, Helfand E (1980) Kinetics of conformational transitions in chain molecules. J Chem Phys 72: 5489-5500
Stigter D (1974a) Micelle formation by ionic surfactants. I. Two phase model, Gouy-Chapman model, hydrophobic interactions. J Colloid Interface Sci 47: 473-482
Stigter D (1974b) Micelle formation by ionic surfactants. II. Specifity of head groups, micelle structure. J Phys Chem 78: 2480-2485
Szajdzinska-Pietek E, Maldonado R, Kevan L, Jones RRM (1984) Electron spin resonance and electron spin echo modulation studies of N,N,N',N'-

tetramethylbenzidine photoionization in anionic micelles: structural effects of tetramethylammonium cation counterion substitution for sodium cation in dodecyl sulfate micelles. J Am Chem Soc 106: 4675-4678
Szajdzinska-Pietek E, Maldonado R, Kevan L, Jones RRM (1985) Electron spin resonance and electron spin echo modulation studies of N,N,N',N'-tetramethylbenzidine photoionization in frozen micellar solutions: structural effect of 1-butanol addition to sodium and tetramethylammonium dodecylsulfate and dodecyl trimethylammonium chloride micelles. J Am Chem Soc 107: 6467-6470
Tanford C (1980) The hydrophobic effect: formation of micelles and biological membranes. Wiley Interscience, New York
Thomas JK, Grieser F, Wong M (1978) Fast reactions in micelles. Ber Bunsenges Phys Chem 82: 937-949
Turro NJ, Grätzel M, Braun AM (1980) Photophysical and photochemical processes in micellar systems. Angew Chem 92: 712-734
Wallace SC, Grätzel M, Thomas JK (1973) Laser photoionization of aromatic hydrocarbons in micellar solutions. Chem Phys Lett 23: 359-362
Zana R, Yiv S, Strazielle C, Lianos P (1980) Effect of alcohol on the properties of micellar systems. I. Critical micellization concentration, micelle molecular weight and ionization degree, and solubility of alcohols in micellar solutions. J Colloid Interface Sci 80: 208-223

6. MOLECULAR ORIGINS OF REACTIVITY DISTRIBUTION IN CONDENSED MEDIA

It seems that with the use of the time-dependent rate constant of the form

$$k(t) = Bt^{\alpha-1} , \qquad 0 < \alpha \leq 1 \qquad (6.1)$$

there was achieved the reaction kinetics of outstanding universality. It takes into account the distribution of reactivity of species trapped in condensed media and therefore provides an adequate description of reaction patterns for a variety of systems so diverse as those presented in Chapters 1 - 5. Adequate description of the reaction patterns has to be followed by the development of the proper reaction models. Here appears the problem of the molecular origins of reactivity distributions in condensed media. Some remarks were given earlier, cf. the distribution of reactivity in the tunnelling models for reactions of trapped electrons in alkaline glasses, or the effect of cooperative movement of surfactant molecules on the reactivity of species embedded in micelles. They indicated much more specific origins than the vague interpretation in terms of a distribution of environments or trapping sites which in many cases has more formal than physical significance. Strong arguments were given, cf. the discussion of reaction mechanism for trapped hydrogen atoms, that in some cases the species are not leaving the intact traps either classically or quantum-mechanically, but are freed to react due to trap destruction. In these cases the distribution of reactivity follows the distribution of trap stability. Here the recent advances of the CTRW are relevant.

Historically, the time-dependent rate constant (6.1) was derived from the continuous time random walk model developed for the anomalous electron transit-time dispersion in amorphous solids (Scher and Montroll 1975) and used to describe the decay of excess electrons trapped in rigid glassy matrices (Hamill and Funabashi 1977). It will be instructive for further discussion to look at the derivation of k(t) from the CTRW model.

TIME-DEPENDENT RATE CONSTANT IN CTRW MODEL

In 1979 Funabashi and Hamill presented detailed derivation of the time-dependent rate constant, eqn. (6.1), according to the CTRW model. The system consists of trapped electrons and acceptors whose concentration is far in excess of that of electrons. The reaction is assumed to be diffusion controlled and the number of electrons arriving at acceptors per unit time are calculated. Mathematically this was equivalent to the calculation of flux of walkers, acceptors, arriving at the origin, electron, for the first time. Since the position of the electron is fixed, the diffusional motion of acceptors corresponds practically to that of the electron itself. It was assumed that each point of the lattice, taken arbitrarily as cubic, is initially occupied by an acceptor. Thus the flux of acceptors arriving at the origin for the first time, I(t), is related to the time-dependent rate constant for scavenging, k(t), defined as the same flux for unit concentration of acceptors, by

$$I(t) = a^{-3}k(t) \tag{6.2}$$

where a is the lattice constant. The number of trapped electrons at time t, n(t), is given by first-order equation

$$n(t)/n(0) = \exp\left[-c_s \int_0^t k(t)dt\right] \tag{6.3}$$

where n(0) is the initial number of trapped electrons and c_s is the concentration of acceptors. The flux I(t) was calculated in the following way. For constant jump time, τ, for each step

$$I(t) = \sum_{n,s} F_n(s)\,\delta(t - n\tau) \ , \tag{6.4}$$

where $F_n(s)$ is the probability that the acceptor reaches the origin for the first-time in n steps starting from the lattice point s. The delta function in (6.4) indicates that acceptors reach the origin only at times which are integral multiples of the jump time τ. To generalize the calculation to the case of multiple-jump times the following generating functions were introduced

$$F(s,z) = \sum_{n=1}^{\infty} F_n(s)z^n \ , \tag{6.5}$$

and

$$P(s,z) = \sum_{n=0}^{\infty} P_n(s)z^n \ , \tag{6.6}$$

where $P_n(s)$ is the probability that the acceptor is at the origin after n steps, not necessarily for the first time, starting from the lattice point s. For these generating functions the following equalities are satisfied

$$F(s,z) = \left[P(s,z) - \delta_{s0}\right]/P(0,z) \tag{6.7}$$

and

$$\sum_s P(s,z) = (1 - z)^{-1} \tag{6.8}$$

For single jump time τ, eqn. (6.4), leads to the same result as that of the usual diffusion treatment, I(t) becomes time-independent after a few steps. It was assumed, however, that at each lattice point the walker has a choice of jump time from a set of jump times, τ_1, τ_2, ..., with corresponding probabilities, C_1, C_2, ... normalized as

$$\sum_i C_i = 1 \, . \tag{6.9}$$

In this case the flux is given by

$$I(t) = \sum_{n,s} \frac{n! \prod_i C_i^{n_i}}{\prod_i n_i!} F_n(s)\, \delta(t - \sum_i n_i \tau_i) \tag{6.10}$$

where

$$\sum_i n_i = n \, . \tag{6.11}$$

Using the relationships

$$\sum_{\{n_i\}} \frac{n! \prod C_i^{n_i}}{\prod n_i} = (\sum_i C_i)^n = 1 \tag{6.12}$$

and

$$\delta(t - \sum_i n_i \tau_i) = \frac{1}{2\pi i} \int_{C-i\infty}^{C+i\infty} \exp\left[u(t - \sum_i n_i \tau_i)\right] du \tag{6.13}$$

equation (6.10) was written as

$$I(t) = \frac{1}{2\pi i} \sum_{n,s} F_n(s) \int_{C-i\infty}^{C+i\infty} \left[\sum_i C_i \exp(-u\tau_i)\right]^n \exp(ut)\, du \tag{6.14}$$

Now it was assumed that the jump time has a continuous spectrum, thus

$$\sum_i C_i \exp(-u\tau_i) \longrightarrow \int_0^\infty \Psi(\tau) \exp(-u\tau)\, d\tau \tag{6.15}$$

where $\Psi(\tau)$ denotes the continuous jump time distribution function. The rhs of (6.15) is the Laplace transform of $\Psi(\tau)$

$$\tilde{\Psi}(u) = \int_0^\infty \Psi(\tau) \exp(-u\tau)\, d\tau \tag{6.16}$$

Using eqns. (6.5), (6.14), (6.15), and (6.16) the flux I(t) was given by

$$I(t) = \sum_s \frac{1}{2\pi i} \int F(s, \Psi(u)) \exp(ut)\, du \tag{6.17}$$

which is equivalent to

$$\tilde{I}(u) = \int_0^{\infty} I(t)\exp(-ut)dt = \sum_s F(s, \tilde{\Psi}(u)) =$$

$$= \frac{1}{\left[1 - \tilde{\Psi}(u)\right] P(0, \tilde{\Psi}(u))} - 1 \qquad (6.18)$$

where in the last step eqns. (6.7) and (6.8) were used.

For the asymptotic form of $\Psi(t)$ introduced by Scher and Montroll (1975)

$$\Psi(t) \sim \left[At^{1+\alpha}\Gamma(1-\alpha)\right]^{-1} \qquad (6.19)$$

where Γ denotes the gamma function, A and $0 < \alpha \leqslant 1$ are constant parameters, the corresponding asymptotic form for $\tilde{\Psi}(u)$ is given by

$$\tilde{\Psi}(u) \sim 1 - n^{\alpha}/A \ , \qquad z \longrightarrow 1 \qquad (6.20)$$

and

$$P(0,z) \longrightarrow 1.516 \ , \qquad z \longrightarrow 1 \qquad (6.21)$$

$$I(u) \sim \beta A u^{-\alpha} \ , \qquad \beta = 0.659 \qquad (6.22)$$

The time-dependent flux has the form

$$I(t) \sim \frac{\beta A t^{\alpha-1}}{\Gamma(\alpha)} \qquad (6.23)$$

and the time-dependent scavenging rate constant is given by

$$k(t) \sim a^3 \beta A t^{\alpha-1}/\Gamma(\alpha) \equiv Bt^{\alpha-1} \qquad (6.24)$$

i.e. by eqn. (6.1).

There were some objections of Shlesinger (1979) to the above derivation but neither the form (6.1) nor eqn. (6.18), we are going to refer to, were questioned. For the model the main question was the physical justification for the application of the class of hoping-time distribution functions with long tails which decay asymptotically as $t \rightarrow \infty$ in the manner given by (6.19). The first justification given for the use of eqn. (6.19) was that the time dependence of $\Psi(t)$ calculated for a spectrum of rate constants appropriate for a random medium is simulated by $t^{-(1+\alpha)}$ (Scher and Montroll 1975). Then there was given (Montroll and Shlesinger 1983) a tale of tails: from maximum entropy formalism through fractals and scaling phenomena to 1/f noise. However, more serious question arise when a time-dependent rate constant was successfully applied for a number of systems discussed above in which hopping of trapped species was hard to imagine. Attention was paid (Plonka et al. 1982) to Kohlrausch (1863) relaxation function used over the years to describe adequately a variety of relaxation phenomena, cf. Rajagopal et al. (1983), in condensed phases. From Kohlrausch relaxation function the time-dependent rate constant (6.1) was deduced, cf. above, and the trapped species were assumed to be freed due to trap destruction caused by matrix relaxation. Then the CTRW model was developed for the nonexponential relaxation.

CTRW MODEL OF NONEXPONENTIAL RELAXATION

CTRW have been recently applied to slow nonexponential relaxation near and below the polymer glass transition. Shlesinger and Montroll (1984) extended Glarum's defect diffusion model of dielectric relaxation to include the possibility that the progress of a defect toward a frozen dipole might be intermittent, consisting of bursts of hopping events with intervening clusters of pauses. To gain the physical picture, one supposes that an electric field has been applied for some time to a medium containing many polar molecules, or polar groups in complex molecules such as a polymer and that the medium has relaxed around the polar groups to the degree that the dipole moments have been frozen in direction after the field has been removed. One has to suppose also that the medium contains defects, which through thermal excitation become mobile, some reaching the location of a frozen dipole and upon doing so relax the medium in the neighbourhood of the dipole to the degree that the dipole may reorient itself as required in an approach to equilibrium. The rate of relaxation is governed by the flux of defects into dipoles. Intermittency results if defects get stuck, either because they wander down dead ends of the polymer chain and have to find their way back or because they must wait to jump over a distribution of barriers. The Laplace transform of the defect flux I(u) is given by eqn. (6.18). In the case of a waiting time density $\Psi(t)$ with a long-tail as (6.19) eqn. (6.18) yields asymptotically for the dipole orientation survival probability $\Phi(t)$

$$\Phi(t) \sim \exp(-ct^{\alpha}) \qquad (6.25)$$

where the constant c depends on defect density, lattice structure and the precise form of $\Psi(t)$. Eqn. (6.25), known as Williams-Watts relaxation function for dielectric relaxation, cf. Montroll and Bendler (1984), is exactly the Kohlrausch relaxation function which for mechanical relaxation can be interpreted similarly, e.g. as the survival probability of a frozen state waiting for the arrival of a defect, cf. Bendler (1984). This kind of mechanism is worth considering for trap destruction in all cases when it is the most likely way to free the trapped species. The required long-time behaviour of $\Psi(t)$, eqn. (6.19), was deduced by Bendler and Shlesinger (1985) from the barrier dispersion model.

BARRIER DISPERSION MODEL OF WAITING-TIME DISTRIBUTION FUNCTION

The phenomenological Eyring (1935) or Kramers (1940) form of a thermally activated jump rate λ over a single free energy barrier F_0^+ is

$$\lambda_0 = \lambda_\infty \exp(-F_0^+/kT) \qquad (6.26)$$

where λ_∞ is the attempt frequency which depends on the potential curvatures and the effective viscosity. Free energy of the barrier is composed of both an entropy and an energy contribution

$$F^+ = -TS^+ + E^+ = -T(S_o + \delta S) + (E_o + \delta E) \tag{6.27}$$

where δS and δE are entropy and energy fluctuations generated by local disorder. If the time scale of the barrier fluctuations is not much shorter than the defect waiting times, a distribution of the latter results. The simplest is the case of linear relation between energy and entropy, $\delta S = \sigma \delta E$. An exponential density of δE is introduced

$$g(\delta E) = q \exp(-q \delta E) \;, \quad \delta E > 0 \;. \tag{6.28}$$

For a single barrier with no memory between hops one has a Poison density of waiting times

$$\Psi(t) = \lambda_o \exp(-\lambda_o t) \;, \tag{6.29}$$

while for a distribution of barriers

$$\Psi(t) = \int_0^\infty \lambda \rho(\lambda) \exp(-\lambda t) d\lambda \tag{6.30}$$

with the density of rates given by

$$\rho(\lambda) = g(\delta E) |d\delta E / d\lambda| \;. \tag{6.31}$$

To calculate $\rho(\lambda)$ one rewrites eqn. (6.27) as

$$F^+ = F_o^+ - T\delta S + \delta E = F_o^+ + \delta E(1 - \sigma T) \tag{6.32}$$

then

$$\lambda/\lambda_o = \exp\left[(F_o^+ - F^+)/kT\right] = \exp\left[\delta E(\sigma T - 1)/kT\right] \tag{6.33}$$

from which

$$\delta E = \ln(\lambda/\lambda_o) kT/(\sigma T - 1) \;. \tag{6.34}$$

Substituting (6.34) into (6.28) one gets

$$g(\delta E) = q \exp\left[q \frac{\ln(\lambda/\lambda_o) kT}{(1 - \sigma T)}\right] = q\left(\frac{\lambda}{\lambda_o}\right)^{\frac{qkT}{1- T}} \;. \tag{6.35}$$

Then

$$\rho(\lambda) = \frac{qkT}{1 - \sigma T} \frac{1}{\lambda} \left(\frac{\lambda}{\lambda_o}\right)^{\frac{qkT}{1- T}} \;, \tag{6.36}$$

or

$$\rho(\lambda) = \frac{\alpha}{\lambda}\left(\frac{\lambda}{\lambda_o}\right)^{\alpha} \tag{6.37}$$

with

$$qkT/(1 - \sigma T) \equiv \alpha . \tag{6.38}$$

Introducing eqn. (6.37) into eqn. (6.30) one gets

$$\psi_{\alpha}(t) = (\alpha/\lambda_o^{\alpha}) \int_0^{\lambda_o} \lambda^{\alpha} \exp(-\lambda t)\, d\lambda \tag{6.39}$$

for $0 \leq \lambda \leq \lambda_o$.

Equation (6.39) can be rewritten as an incomplete γ function

$$\psi_{\alpha}(t) = (\alpha/\lambda_o^{\alpha}) t^{-1-\alpha} \int_0^{\lambda_o t} y^{\alpha} \exp(-y)\, dy =$$

$$= (\alpha/\lambda_o^{\alpha}) t^{-1-\alpha} \gamma(\alpha+1, \lambda_o t) \tag{6.40}$$

which for $t \to \infty$ yields

$$\psi_{\alpha}(t) \sim \left[\alpha \Gamma(1+\alpha)/\lambda_o^{\alpha}\right] t^{-1-\alpha} \tag{6.41}$$

i.e. the required long-time behaviour of $\psi(t)$. As the equivalence of CTRW model and multiple-trapping model was stressed (Noolandi 1977) it is worth noting here the derivation of the waiting-time distribution function with the long-time behaviour similar to (6.41) from the multiple-trapping model by Arkhipov and Rudenko (1982) for exponential trap energy distribution.

Thus, barrier dispersion leads to the long-time behaviour of the waiting-time distribution function given by eqn. (6.19) which in turn, through eqn. (6.18), yields the relaxation function (6.25) i.e. the first-order kinetic equation with time-dependent rate constant (6.1). Now let us recall that the time-dependent rate constant calculated from the tunnelling models, Chapter 2, was found to be well approximated by eqn. (6.1). In the tunnelling models there is also a distribution of energy barriers arising from the distributions of electron - acceptor distances and electron energy in the traps. Thus we may say, cf. Introduction, that time-dependent rate constant (6.1) reflects the distribution of energy barriers for reactions of trapped species in condensed media. However, an energy barrier for reaction in condensed phases, crossed classically or quantum-mechanically, is not necessarily the energy needed for species to leave the intact traps. It may be in many cases the energy needed to destroy the traps by the cooperative movement of the molecules in the close environment or the energy of mobile defect transport.

REFERENCES

Arkhipov VI, Rudenko AI (1982) Drift and diffusion in materials with traps. II. Non-equilibrium transport regime. Phil Mag B 45: 189-207

Bendler JT (1984) Lévy (stable) probability densities and mechanical relaxation in solid polymers. J Statist Phys 34: 625-637

Bendler JT, Shlesinger MF (1985) Derivation of the Kohlrausch-Williams/ Watts decay law from activation energy dispersion. Macromolecules 18: 591-592

Eyring H (1935) The activated complex and the absolute rate of chemical reactions. Chem Rev 17: 65-77

Funabashi K, Hamill WH (1979) The role of electron - phonon interaction and non-Gaussian transport in spectral changes of trapped electrons in glasses. Can J Chem 57: 197-206

Hamill WH, Funabashi K (1977) Kinetics of electron trapping reactions in amorphous solids; a non-Gaussian diffusion model. Phys Rev B 16: 5523-5527

Kohlrausch F (1863) Ueber die elastische Nachwirkung bei der Torsion. Pogg Ann 119: 337-368

Kramers HA (1940) Brownian motion in a field of force and the diffusion model of chemical reactions. Physica 7: 284-304

Montroll EW, Bendler JT (1984) On Lévy (or stable) distributions and the Williams-Watts model of dielectric relaxation. J Statist Phys 34: 129-162

Montroll EW, Shlesinger MF (1983) Maximum entropy formalism, fractals, scaling phenomena, and 1/f noise: a tale of tails. J Statist Phys 32: 209-230

Noolandi J (1977) Equivalence of multiple-trapping model and time-dependent random walk. Phys Rev B 16: 4474-4479

Plonka A, Bogus W, Kroh J (1982) Effect of irradiation temperature on the decay rate of trapped hydrogen atoms in sulfuric acid glasses. Radiat Phys Chem 19: 389-392

Rajagopal K, Ngai KL, Rendell RW, Teitler S (1983) Nonexponential decay in relaxation phenomena. J Statist Phys 30:285-292

Scher H, Montroll EW (1975) Anomalous transit-time dispersion in amorphous solids. Phys Rev B 12: 2455-2477

Shlesinger MF (1979) Electron scavenging in glasses. J Phys Chem 70: 4813-4818

Shlesinger MF, Montroll EW (1984) On the Williams-Watts function of dielectric relaxation. Proc Natl Acad Sci US 81: 1280-1283

7. APPENDICES

A. KINETICS WITH TIME-DEPENDENT RATE CONSTANT

Let the bimolecular reaction

$$A + B \longrightarrow AB \tag{A.1}$$

be described by a second-order kinetic equation

$$-dc_A/dt = k(t)c_A c_B \tag{A.2}$$

with time-dependent rate constant

$$k(t) = Bt^{\alpha-1} , \qquad 0 < \alpha \leqslant 1 \tag{A.3}$$

In the absence of other processes, conservation of particles implies

$$c_A + c = c_B \tag{A.4}$$

where c is a constant. To integrate (A.2) one can write

$$-\int_{c_A(0)}^{c_A} \frac{1}{c_A(c_A + c)} dc_A = \int_0^t k(t)dt \tag{A.5}$$

or

$$\frac{1}{c}\int_{c_A(0)}^{c_A}\left(\frac{1}{c_A} - \frac{1}{c_A + c}\right)dc_A = -B\int_0^t t^{\alpha-1}dt \tag{A.6}$$

and get

$$\frac{1}{c}\left\{\ln \frac{c_A}{c_A(0)} - \ln \frac{c_A + c}{c_A(0) + c}\right\} = -Bt^{\alpha}/\alpha . \tag{A.7}$$

For $c \neq 0$ eqn. (A.7) can be rewritten as

$$\left[c_A/(c_A + c)\right]/\left[c_A(0)/(c_A(0) + c)\right] = \exp(-cBt^{\alpha}/\alpha) . \tag{A.8}$$

Two limiting cases are of interest. If $c_B(0) \gg c_A(0)$ then $c_A + c \approx \approx c_A(0) + c$ and eqn. (A.8) yields

$$c_A/c_A(0) = \exp(-cBt^{\alpha}/\alpha) \tag{A.9}$$

which is proper for the pseudo first-order decay. If $c_A(0) = c_B(0)$ then for $c \rightarrow 0^+$, cf. Blumen et al. (1983)

$$\frac{1}{c}\ln(1 + \frac{c}{c_A}) \sim \frac{1}{c_A} \tag{A.10}$$

and (A.6) yields

$$c_A^{-1} - c_A(0)^{-1} = Bt^{\alpha}/\alpha \tag{A.11}$$

for second-order equal concentration kinetics.

The first-order kinetic equation

$$-dc/dt = k(t)c \tag{A.12}$$

with time-dependent rate constant (A.3) yields upon integration the relation

$$c/c(0) = \exp(-Bt^{\alpha}/\alpha) \tag{A.13}$$

which can be interpreted as superposition of many simple exponential decays, i.e.

$$\exp\left[-(t/\tau_0)^{\alpha}\right] = \int_0^{\infty} f(\tau)\exp(-t/\tau)d\tau \tag{A.14}$$

for the effective lifetime τ_0 equal to

$$\tau_0 = (\alpha/B)^{1/\alpha} \quad . \tag{A.15}$$

This interpretation, giving the distribution of reactivity $f(\tau)$, cf. Appendix B, seems to be proper also for the pseudo first-order decay, eqn. (A.9), and extendible for second-order unequal concentration kinetics, eqn. (A.8). It cannot be applied for the second-order equal concentration kinetics. At the moment, this is rather depressing as this kind of kinetics is very popular.

Temperature dependence of k(t) given by eqn. (A.3) is contained in those of B and α, or according to eqn. (A.15) in τ_0 and α. In the case of first-order decay, to apply the Arrhenius equation it is convenient to use the mean lifetime $\bar{\tau}$ deduced in the following way. If there was c(0) species initially, there remains at time t the number c given by eqn. (A.13). At this very time t the number decaying during an interval of time dt is

$$k(t)c\,dt \ , \quad \text{or}$$

$$Bt^{\alpha-1}c(0)\exp(-Bt^{\alpha}/\alpha)dt \ ,$$

and since each of them has existed for time t, the total lifetime is

$$Bt^{\alpha-1}c(0)\exp(-Bt^{\alpha}/\alpha)tdt \; .$$

Therefore, the mean lifetime $\bar{\tau}$ is

$$\bar{\tau} = (1/c(0)) \int_0^{\infty} c(0)Bt^{\alpha} \exp(-Bt^{\alpha}/\alpha)dt =$$

$$= (\alpha/B)^{1/\alpha}\Gamma(1 + 1/\alpha) \; , \quad \text{or}$$

$$= (\tau_0/\alpha)\,\Gamma(1/\alpha) \; , \qquad (A.16)$$

where Γ denotes the gamma function.

In the case of second-order equal concentration kinetics, to apply the Arrhenius equation one is forced to use the half-life $\tau_{1/2}$ which according to eqn. (A.11) is equal to

$$\tau_{1/2} = (\alpha/c(0)B)^{1/\alpha} \; , \quad \text{or}$$

$$= \tau_0 c(0)^{1/\alpha} \; . \qquad (A.17)$$

B. INVERSE LAPLACE TRANSFORM OF KOHLRAUSCH RELAXATION FUNCTION

If we regard

$$c/c(0) = \exp\left[-(t/\tau_0)^{\alpha}\right] \qquad (B.1)$$

as the superposition of many simple exponential decays

$$\exp\left[-(t/\tau_0)^{\alpha}\right] = \int_0^{\infty} f(\tau)\exp(-t/\tau)d\tau \qquad (B.2)$$

the distribution function $f(\tau)$ is given by an inverse Laplace transform of the lhs of eqn. (B.2). There seems to be no closed-form analytical solution for α from the range (0,1). The result evaluated by saddle point calculation for large τ is (Majumdar 1971; Scher and Montroll 1975)

$$f(\tau) \sim \frac{(\alpha\tau)^{(2-\alpha)/2(1-\alpha)}}{\tau^2\left[2\pi\alpha(1-\alpha)\right]^{1/2}\tau_0^{\alpha/2(1-\alpha)}} \cdot$$

$$\cdot \exp\left[-(1-\alpha)(\alpha\tau/\tau_0)^{\alpha/(1-\alpha)}\right] \; . \qquad (B.3)$$

This general asymptotic form (B.3) is exactly equal to the closed-form analytical results known for $\alpha = 1/2$ (Scher and Montroll 1975) i.e. to

$$f(\tau) = \frac{1}{2}\tau_0^{-1}(\tau_0/\pi\tau)^{1/2}\exp(-\tau/4\tau_0) \qquad (B.4)$$

There is also known the closed-form analytical solution for $\alpha = 1/3$ (Scher and Montroll 1975)

$$f(\tau) = \frac{\sin(\pi/3)}{\pi\tau} xK_{1/3}(x) \tag{B.5}$$

where $K_\nu(x)$ is the modified Bessel function of order ν of

$$x \equiv 2\cdot 3^{-3/2}(\tau_o/\tau)^{-1/2} \tag{B.6}$$

For $x \longrightarrow \infty$

$$xK_{1/3}(x) \longrightarrow (\pi x/2)^{1/2}\exp(-x) \tag{B.7}$$

and (B.5) yields

$$f(\tau) = (2\cdot 3^{1/4}\pi^{1/2}\tau^{3/4}\tau_o^{1/4})^{-1}\exp\left[-2\cdot 3^{-3/2}(\tau_o/\tau)^{-1/2}\right] \tag{B.8}$$

i.e. exactly what one gets from (B.3) for $\alpha = 1/3$.

For $\alpha \longrightarrow 0$ (Montroll and Bendler 1984)

$$f(\tau) = \alpha\tau_o^{-1}(\tau_o/\tau)^{1-\alpha}\exp\left[-(\tau/\tau_o)^{\alpha}\right] , \tag{B.9}$$

and for small but increasing α the proper correction terms are

$$\begin{aligned} f(\tau) &= \alpha\tau_o^{-1}(\tau_o/\tau)^{1-\alpha}\exp\left[-(\tau/\tau_o)^{\alpha}\right]\cdot \\ &\cdot\Big[U_1 - \alpha U_2(1-\mu^{\alpha}) + \alpha^2 U_3(1 - 3\mu^{\alpha} + \mu^{2\alpha}) + \\ &- \alpha^3 U_4(1 - 7\mu^{\alpha} + 6\mu^{2\alpha} - \mu^{3\alpha}) + \\ &+ \alpha^4 U_5(1 - 15\mu^{\alpha} + 25\mu^{2\alpha} - 10\mu^{3\alpha} + \mu^{4\alpha})\ldots\Big] \end{aligned} \tag{B.10}$$

where $\mu \equiv \tau/\tau_o$, and

$$U_1 = 1$$

$$U_2 = \gamma = 0.577215665$$

$$U_3 = \frac{1}{2}\gamma^2 - \frac{\pi^2}{12} = -0.65587807$$

$$U_4 = \frac{\gamma^3}{6} + \frac{1}{3}\zeta(3) - \frac{\gamma\pi^2}{12} = -0.04200264$$

$$U_5 = \frac{\gamma^4}{24} - \frac{\pi^2\gamma^2}{24} + \frac{\pi^4}{1440} + \frac{\gamma\zeta(3)}{3} = 0.16653861$$

with Euler's gamma

$$\gamma = 0.5772156649$$

and Riemann's zeta function $\zeta(m)$

$$\zeta(3) = 1.2020569031 \; .$$

To illustrate the applicability of saddle point approximation and series expansion the distribution densities calculated according to eqns. (B.3), (B.5), and (B.10) are given in Table B.1. Looking at the numerical values presented in Table B.1 one might be concerned with the singularity of $f(\tau)$ as $\tau \rightarrow 0$, which implies the density $f(\tau)$ to be infinite at the origin. Actually, this causes no difficulty (Montroll and Bendler 1984) since the singularity is very weak. The quantity that gives a better description of the distribution of reactivity near the origin is the cummulative distribution function

$$P(\tau) = \int_0^{\tau} f(\tau)\,d\tau \tag{B.11}$$

which gives the fraction of $f(\tau)$ with lifetimes less than τ. By direct integration of $f(\tau)$ for small α, eqn. (B.9), one gets

$$P(\tau) = (1/\alpha\pi)\,\Gamma(\sin\pi\alpha)(\tau/\tau_0)^{\alpha} \tag{B.12}$$

which vanishes at the origin. There is no serious accumulation of $f(\tau)$ until τ is a significant fraction of τ_0.

Table B.1. Comparison of distribution densities $f(\tau)$ for $\alpha = 1/3$ calculated according to eqns. (B.3), (B.5), and (B.10) for $\tau_0 = 1$

x	Exact Eqn. (B.5)	τ	Approximation by Eqn. (B.10)	Eqn. (B.3)
0.1	1.18427	0.0675	1.18462	1.46456
0.2	0.40418	0.2700	0.40402	0.46852
0.3	0.29543	0.6075	0.20521	0.23076
0.4	0.12310	1.0800	0.12291	0.13562
0.5	0.08078	1.6875	0.08063	0.08781
0.6	0.05616	2.4300	0.05605	0.06044
0.7	0.04063	3.3075	0.04056	0.04340
0.8	0.03028	4.3200	0.03024	0.03214
0.9	0.02308	5.4675	0.02306	0.02437
1.0	0.01790	6.7500	0.01791	0.01883
1.5	0.00599	15.1875	0.00602	0.00622
2.0	0.00238	27.0000	0.00241	0.00245
2.5	0.00104	42.1875	0.00105	0.00106

Sometimes it is convenient to discuss the distribution of reactivities in terms of rate constant distribution f(k) with

$$k = 1/\tau \ . \qquad (B.13)$$

Here the following relations hold

$$f(k) = f(\tau)\,|d\tau/dk| = f(\tau)\tau^2 \ , \qquad (B.14)$$

or

$$f(k)k = f(\tau)\tau \ . \qquad (B.15)$$

Before leaving this subject it is worth to look at the series expansion of $f(\tau)$ in α for small α (Montroll and Bendler 1984)

$$\begin{aligned} f(\tau) = (\alpha/e\tau)\left\{\exp\left[-\tfrac{1}{2}\alpha^2\ln^2(\tau/\tau_o)\right]\right\}\cdot \\ \cdot\left\{1 - \alpha^2\left[U_3 - U_2\ln(\tau/\tau_o)\right] + \right. \\ \left. + \alpha^3\left[U_4 - U_3\ln(\tau/\tau_o) + \tfrac{1}{2}U_2\ln^2(\tau/\tau_o)\right] \ldots\right\} \end{aligned} \qquad (B.16)$$

to find that for very small values of

$$f(\tau)d\tau = (\alpha/e)\exp\left[-\tfrac{1}{2}\alpha^2\ln^2(\tau/\tau_o)\right] d\,\ln(\tau/\tau_o) \qquad (B.17)$$

$\ln(\tau/\tau_o)$ has a normal distribution. Since mechanisms are known that lead to ln normal distribution Montroll and Bendler (1984) exploit the appearance of that distribution to show the mechanism that may imply the relaxation process adequate to eqn. (B.1) at least in the small α regime. They consider a complex event whose successful occurence requires the successful conclusion of a sequence of n independent other events. The probability of the occurance of the required event in a unit time is

$$P = p_1p_2p_3\ldots p_n \qquad (B.18)$$

where p_j denotes the probability of success of the jth subevent in the required time, and

$$\ln P = \ln p_1 + \ln p_2 + \ln p_3 + \ldots + \ln p_n \qquad (B.19)$$

Since p_j are independent random variables, so are $\ln p_j$. If there exist the appriopriate moments for $\ln p_j$ and n is large, the central limit theorem is applicable and ln P has a normal distribution. In an ensemble of possible events the expected time for its occurance, τ, is proportional to 1/P, and

$$\ln\tau = \ln(1/P) + \text{const} = \sum_{j=1}^{n}\ln(1/p_j) + \text{const} \qquad (B.20)$$

Again, since each p_j is a random variable so is $\ln(1/p_j)$ and if n is large and there exist the proper moments of $\ln(1/p_j) = -\ln p_j$, $\ln\tau$ has a normal distribution. As an example the mechanism of dielectric relaxation is pointed out, cf. Chapter 6. If the mobile defect must pass over many free energy barriers in its approach to a frozen dipole the conditions required for the existence of ln normal distribution are satisfied. An alternative model would be the one requiring the frozen dipole itself to surmount a large number of potential barriers in its attempt to achieve its final equilibrium state. It is a pity that in dielectric relaxation the lowest reported values of α are about 0.2 (Montroll and Bendler 1984) too high for ln normal distribution. From the surveyed data only the post-irradiation decay of trapped electrons do proceed with α as low as 0.1, and indeed the distribution of their reactivities mimics the ln normal distribution.

REFERENCES

Blumen A, Klafter J, Zumofen G (1983) Recombination in amorphous materials as a continuous-time random-walk problem. Phys Rev B 27: 3429-3435

Majumdar CK (1971) Stress relaxation function of glass. Solid State Comm 9: 1087-1090

Montroll EW, Bendler JT (1984) On Lévy (or stable) distributions and the Williams-Watts model of dielectric relaxation. J Statist Phys 34: 129-162

Scher H, Montroll EW (1975) Anomalous transit-time dispersion in amorphous solids. Phys Rev B 12: 2455-2477

Lecture Notes in Chemistry

Vol. 1: G. H. Wagnière, Introduction to Elementary Molecular Orbital Theory and to Semiempirical Methods. V, 109 pages. 1976.

Vol. 2: E. Clementi, Determination of Liquid Water Structure. VI, 107 pages. 1976.

Vol. 3: S. R. Niketic and K. Rasmussen, The Consistent Force Field. IX, 212 pages. 1977.

Vol. 4: A. Graovac, I. Gutman and N. Trinajstić, Topological Approach to the Chemistry of Conjugated Molecules. IX, 123 pages. 1977.

Vol. 5: R. Carbo and J. M. Riera, A General SCF Theory. XII, 210 pages. 1978.

Vol. 6: I. Hargittai, Sulphone Molecular Structures. VIII, 175 pages. 1978.

Vol. 7: Ion Cyclotron Resonance Spectrometry. Edited by H. Hartmann and K.-P. Wanczek. VI, 326 pages. 1978.

Vol. 8: E. E. Nikitin and L. Zülicke, Selected Topics of the Theory of Chemical Elementary Processes. X, 175 pages. 1978.

Vol. 9: A. Julg, Crystals as Giant Molecules. VII, 135 pages. 1978.

Vol. 10: J. Ulstrup, Charge Transfer Processes in Condensed Media. VII, 419 pages. 1979.

Vol. 11: F. A. Gianturco, The Transfer of Molecular Energies by Collision: Recent Quantum Treatments. VIII, 328 pages. 1979.

Vol. 12: The Permutation Group in Physics and Chemistry. Edited by J. Hinze. VI, 230 pages. 1979.

Vol. 13: G. Del Re et al., Electronic States of Molecules and Atom Clusters. VIII, 177 pages. 1980.

Vol. 14: E. W. Thulstrup, Aspects of the Linear and Magnetic Circular Dichroism of Planar Organic Molecules. VI, 100 pages. 1980.

Vol. 15: A.T. Balaban et al, Steric Fit in Quantitative Structure-Activity Relations. VII, 178 pages. 1980.

Vol. 16: P. Čársky and M. Urban, Ab Initio Calculations. VI, 247 pages. 1980.

Vol. 17: H. G. Hertz, Electrochemistry. X, 254 pages. 1980.

Vol. 18: S. G. Christov, Collision Theory and Statistical Theory of Chemical Reactions. XII, 322 pages. 1980.

Vol. 19: E. Clementi, Computational Aspects for Large Chemical Systems. V, 184 pages. 1980.

Vol. 20: B. Fain, Theory of Rate Processes in Condensed Media. VI, 166 pages. 1980.

Vol. 21: K. Varmuza, Pattern Recognition in Chemistry. XI, 217 pages. 1980.

Vol. 22: The Unitary Group for the Evaluation of Electronic Energy Matrix Elements. Edited by J. Hinze. VI, 371 pages. 1981

Vol. 23: D. Britz, Digital Simulation in Electrochemistry. X, 120 pages. 1981.

Vol. 24: H. Primas, Chemistry, Quantum Mechanics and Reductionism. XII, 451 pages. 1981.

Vol. 25: G. P. Arrighini, Intermolecular Forces and Their Evaluation by Perturbation Theory. IX, 243 pages. 1981.

Vol. 26: S. Califano, V. Schettino and N. Neto, Lattice Dynamics of Molecular Crystals. VI, 309 pages. 1981.

Vol. 27: W. Bruns, I. Motoc, and K. F. O'Driscoll, Monte Carlo Applications in Polymer Science. V, 179 pages. 1982.

Vol. 28: G. S. Ezra, Symmetry Properties of Molecules. VIII, 202 pages. 1982.

Vol. 29: N.D. Epiotis, Unified Valence Bond Theory of Electronic Structure. VIII, 305 pages. 1982.

Vol. 30: R.D. Harcourt, Qualitative Valence-Bond Descriptions of Electron-Rich Molecules: Pauling "3-Electron Bonds" and "Increased-Valence" Theory. X, 260 pages. 1982.

Vol. 31: H. Hartmann, K.-P. Wanczek, Ion Cyclotron Resonance Spectrometry II. XV, 538 pages. 1982.

Vol. 32: H.F. Franzen, Second-Order Phase Transitions and the Irreducible Representation of Space Groups. VI. 98 pages. 1982.

Vol. 33: G.A. Martynov, R.R. Salem, Electrical Double Layer at a Metal-dilute Electrolyte Solution Interface. VI, 170 pages. 1983.

Vol. 34: N.D. Epiotis, Unified Valence Bond Theory of Electronic Structure · Applications. VIII, 585 pages. 1983.

Vol. 35: Wavefunctions and Mechanisms from Electron Scattering Processes. Edited by F.A. Gianturco and G. Stefani. IX, 279 pages. 1984.

Vol. 36: J. Ugi, J. Dugundji, R. Kopp and D. Marquarding, Perspectives in Theoretical Stereochemistry. XVII, 247 pages. 1984.

Vol. 37: K. Rasmussen, Potential Energy Functions in Conformational Analysis. XIII, 231 pages. 1985.

Vol. 38: E. Lindholm, L. Åsbrink, Molecular Orbitals and their Energies, Studied by the Semiempirical HAM Method. X, 288 pages. 1985.

Vol. 39: P. Vanýsek, Electrochemistry on Liquid/Liquid Interfaces. 2, 3–106 pages. 1985.

Vol. 40: A. Plonka, Time-Dependent Reactivity of Species in Condensed Media. V, 151 pages. 1986.

GPSR Compliance

The European Union's (EU) General Product Safety Regulation (GPSR) is a set of rules that requires consumer products to be safe and our obligations to ensure this.

If you have any concerns about our products, you can contact us on ProductSafety@springernature.com

In case Publisher is established outside the EU, the EU authorized representative is:

Springer Nature Customer Service Center GmbH
Europaplatz 3
69115 Heidelberg, Germany

Batch number: 10704763

Printed by Printforce, the Netherlands